sona
BOOKS

**First published in the UK 2021 by Sona Books
an imprint of Danann Media Publishing Ltd.**

Copy Editor Juliette O'Neill

CAT NO: **SON0509**
ISBN: **978-1-912918-74-4**

Made in EU.

The Big Book of
The Human Body

The human body is truly an amazing thing. Capable of awe-inspiring feats of speed and agility, while being mind-blowing in complexity, our bodies are unmatched by any other species on Earth. In The Big Book of the Human Body, we explore our amazing anatomy in fine detail before delving into the intricacies of the complex processes, functions and systems that keep us going. For instance, did you know you really have 16 senses? We also explain the weirdest and most wonderful bodily phenomena, from blushing to hiccuping, cramps to jaundice. We will tour the human body from skull to metatarsal, using anatomical illustrations, amazing photography and authoritative explanations to teach you more. This book will help you understand the wonder that is the human body and in no time you will begin to see yourself in a whole new light!

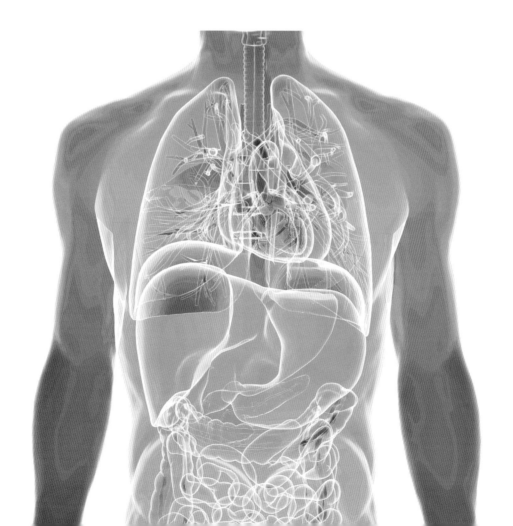

CONTENTS

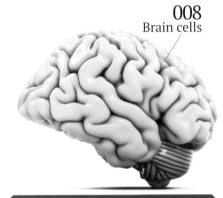
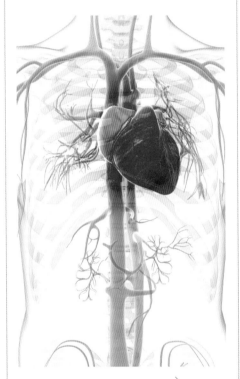

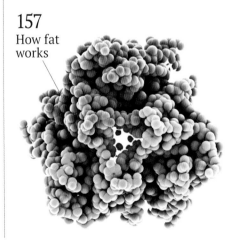

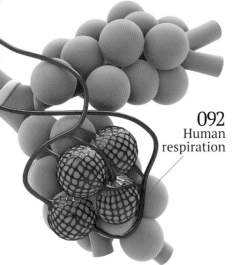

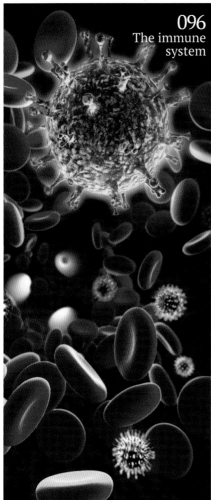

Curious questions

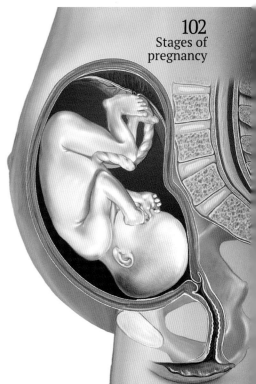

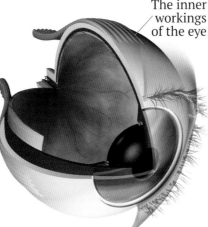

A-Z
of the
HUMAN BODY

Take a tour of your anatomy with
our head-to-toe guide

Alveoli

a As an adult, your lungs have a total surface area of around 50 square metres. That's around a quarter of the size of a tennis court! Packing all of that into your chest is no mean feat, and the body does it using structures called alveoli. They look a little bit like bunches of grapes, packed tightly inside the lungs in order to maximise the use of the available volume in the chest. When you breathe in, they expand, filling with air. The surfaces of the alveoli are just one cell thick and surrounded by tiny blood vessels called capillaries, allowing gases to diffuse easily in and out of the blood with each breath you take.

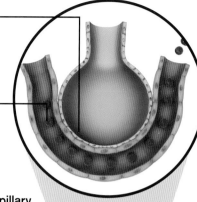

Gas exchange
Gases are swapped at the surface of the alveoli – they travel in or out of the capillary by diffusion.

Red blood cells
Blood cells move through the capillaries in single file, picking up oxygen and dropping carbon dioxide as they go.

Understanding alveoli
How does your body pack such a huge surface area inside your chest?

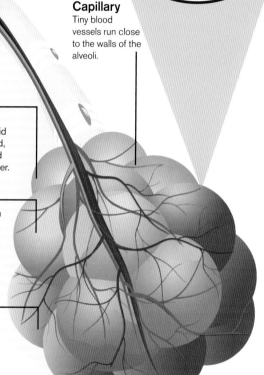

Branching
The lungs are branched like trees, packing as many alveoli as possible into a small space.

Surfactant
Some of the pneumocytes produce a surfactant, a fluid similar to washing-up liquid, which coats the alveoli and stops them sticking together.

Pneumocytes
The alveoli are made from thin, flat cells called pneumocytes, minimising the distance that gases have to travel.

Alveolus
Each individual air sac in the lungs is known as an alveolus.

Capillary
Tiny blood vessels run close to the walls of the alveoli.

Brain

b The brain is not just the most complex structure in the human body, but it is also the most complex object in the known universe. It contains an estimated 86 billion nerve cells, each of which makes hundreds, or even thousands of connections to the others around it.

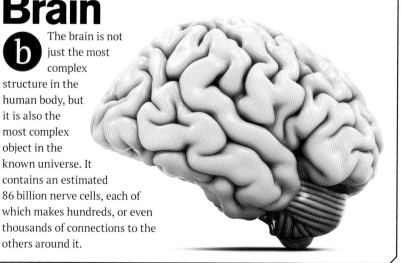

Cornea

c The cornea is the protective coating that keeps your eye free of dust and debris. It looks clear but is actually made up of several layers of cells. Light bends slightly as it passes through the cornea, helping to focus incoming rays on the back of your eye.

It is, in fact, possible to donate corneas for transplant, helping to restore vision to people with corneal damage.

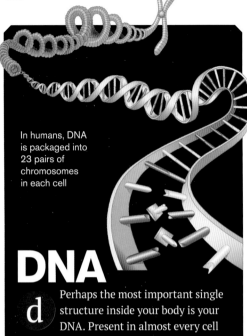

In humans, DNA is packaged into 23 pairs of chromosomes in each cell

DNA

d Perhaps the most important single structure inside your body is your DNA. Present in almost every cell (red blood cells get rid of theirs), it carries the genetic recipes needed to build, grow, repair and maintain you. These recipes are written in combinations of four-letter code (ACTG), and in humans are 3 billion letters long.

Fat

f You have two main types of fat: brown and white. Brown fat burns calories to keep you warm, while white fat stores energy and produces hormones. Children have more brown fat than adults, and it's mainly found in the neck and shoulders, around the organs, and along the spinal cord.

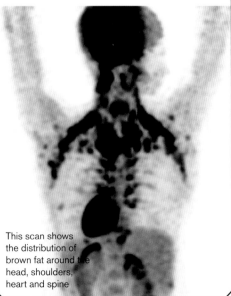

This scan shows the distribution of brown fat around the head, shoulders, heart and spine

Enzymes

e Enzymes are often called 'biological catalysts', and their job is to speed up chemical reactions. You are full of dissolved chemicals with the potential to come together or break apart to form the biological building blocks that you need to stay alive, but the reactions happen too slowly on their own.

Enzymes are molecules with 'active sites' that lock on to other molecules, bringing them close together so that they can react, or bending their structures so that they can combine or break apart more easily. The enzymes themselves do not actually get involved in the reactions; they just help them to happen faster.

Some of the most well-known enzymes are the ones in your digestive system. These are important for breaking down the molecules in your food. However, these aren't the only enzymes in your body. There are others responsible for building molecules, snipping molecules, tidying up when molecules are no longer needed, and even destroying invading pathogens.

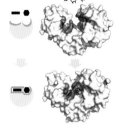

This enzyme brings two molecules close together so that they can react

Digestive enzymes

These microscopic molecules break your food down into absorbable chunks

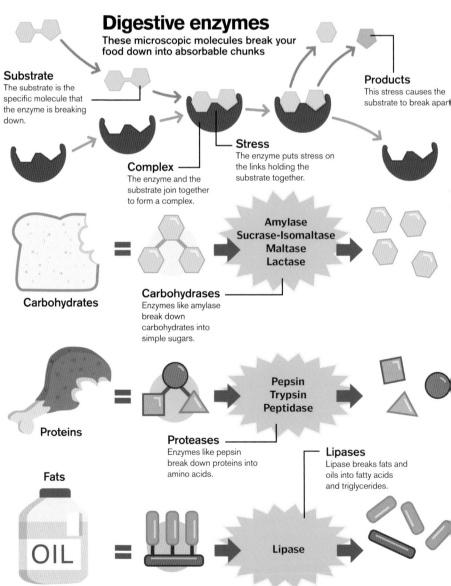

Substrate
The substrate is the specific molecule that the enzyme is breaking down.

Complex
The enzyme and the substrate join together to form a complex.

Stress
The enzyme puts stress on the links holding the substrate together.

Products
This stress causes the substrate to break apart

Carbohydrates

Carbohydrases
Enzymes like amylase break down carbohydrates into simple sugars.

Amylase
Sucrase-Isomaltase
Maltase
Lactase

Proteins

Proteases
Enzymes like pepsin break down proteins into amino acids.

Pepsin
Trypsin
Peptidase

Fats

OIL

Lipase

Lipases
Lipase breaks fats and oils into fatty acids and triglycerides.

Glands

g These structures are responsible for producing and releasing fluids, enzymes and hormones into your body. There are two major types: endocrine and exocrine. Exocrine glands produce substances like sweat, saliva and mucus, and release these through ducts onto the skin or surfaces of other organs. Endocrine glands produce hormones, which are released into the blood to send chemical signals across the body.

The pancreas has both endocrine glands (blue clusters) and exocrine glands (green branches)

As we age, the thickness and colour of our hair changes

Hair

h You have around 5 million hair follicles and, surprisingly, only around 100,000 of those are on your scalp. The others are spread across your body – on your skin, lining your eyelids, and inside your nose and ears. Hair has many functions, helping to keep you warm, trapping dirt and debris, and even (in the case of eyebrows) diverting sweat and rainwater away from your eyes.

Intestines

i After exiting your stomach, food enters your intestines and begins a 7.5-metre journey out of your body. The small intestine comes first, and is filled with digestive enzymes that get to work breaking down and absorbing the molecules from your meal. After this, the large intestine absorbs as much water as possible before the waste is passed out.

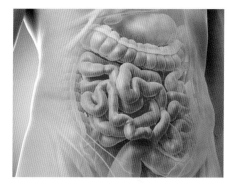

Joints

j There are more than 200 bones in the human body, and to make you move in all the right places, they are linked together by different types of joints.

In your hips and shoulders, you've got ball and socket joints, which allow the widest range of movement. They allow movement forwards, backwards, side-to-side and around in circles.

At the knees and elbows, you have hinge joints, which open and close just like a door. And in your wrists and ankles, there are gliding joints, which allow the bones to flex past one another. In your thumb, there is a saddle joint that enables a side-to-side and open-close motion.

Cartilage covers the ends of the bones at many joints, helping to prevent the surfaces from rubbing together, and cushioning the impact as you move. Many joints are also contained within a fluid-filled capsule, which provides lubrication to keep things moving smoothly. These are called synovial joints.

"There are more than 200 bones in the human body"

Types of joints

Each type of joint in your body allows for a different range of movement

Pivot
These joints are adapted for turning, but they do not allow much side-to-side or forwards and backwards movement.

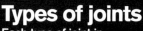

Hinge
The knees and elbows can move forwards and backwards, but not side to side.

Gliding
Gliding joints are found between flat bones, enabling them to slide past one another.

Immovable
Some bones are fused together to form joints that don't actually move, including the bones that make up the skull.

Ball and socket
These joints allow the widest range of movement. The end of one bone is shaped like a ball, and rotates inside another cup-shaped bone.

Saddle
The only saddle joints in the human body are in the thumbs. They allow forwards, backwards and sideways motion, but only limited rotation.

Ellipsoidal
These joints, such as at the base of your index finger, allow forward and backwards movement, and some side-to-side, but they don't rotate.

Kidneys

k Your kidneys keep your blood clean and your body properly hydrated. Blood passes in through knots of blood vessels that are wider on the way in and narrower on the way out. This creates an area of high pressure that forces water and waste out through gaps in the vessel walls. Blood cells and proteins remain in the bloodstream. Each kidney has around a million of these miniature filtering systems, called nephrons, cleaning the blood every time it passes through.

The fluid then tracks through bendy tubes (known as convoluted tubules), where important minerals are collected and returned to the blood. Excess water and waste products are sent on to the bladder as urine to be excreted. Depending on how much salt and water are in your body, your kidneys adjust the amount of fluid that they get rid of, helping to keep your hydration levels stable.

"Your kidneys keep your blood clean and your body hydrated"

The kidneys

These simple-looking organs are packed with microscopic filtration machinery

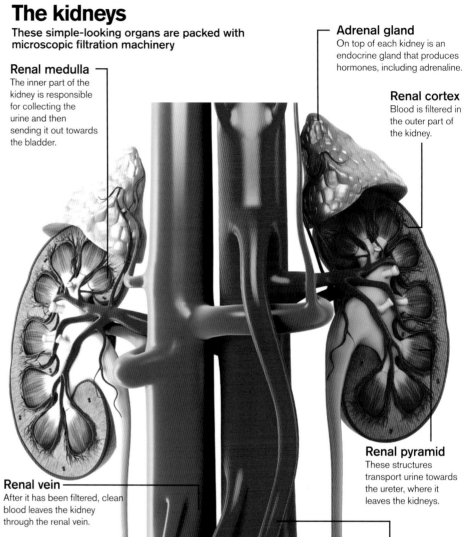

Renal medulla
The inner part of the kidney is responsible for collecting the urine and then sending it out towards the bladder.

Adrenal gland
On top of each kidney is an endocrine gland that produces hormones, including adrenaline.

Renal cortex
Blood is filtered in the outer part of the kidney.

Renal pyramid
These structures transport urine towards the ureter, where it leaves the kidneys.

Renal vein
After it has been filtered, clean blood leaves the kidney through the renal vein.

Ureter
Urine produced by the kidneys travels to the bladder for storage.

Renal artery
Blood enters the kidney through the renal artery.

Lymphatic system

l Everyone knows about the circulatory system that transports blood around the body, but there is a second network of tubes and vessels that is often forgotten. The lymphatic system collects fluid from the tissues, and returns it to the blood via veins in the chest. It is also used by the immune system to monitor and fight infection.

The lymphatic system is studded with lymph nodes, used as outposts by the immune system

Mitochondria

m We know that our bodies need oxygen and nutrients to survive, and mitochondria are the powerhouses that turn these raw materials into energy. There are hundreds in every cell, and they use a complex chain of proteins that shuffle electrons around to produce chemical energy in a form that can be easily used.

Mitochondria have a distinctive two-layered structure, with folds inside

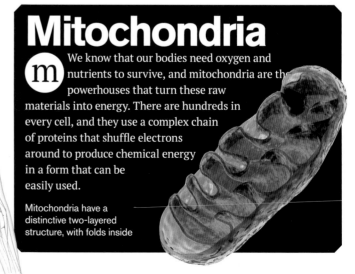

Nervous system

n This is your body's electrical wiring, transmitting signals from your head to your toes and everywhere in between. The nervous system can be split into two main parts: central and peripheral.

The central nervous system is the brain and spinal cord, and makes up the control centre of your body. While the brain is in charge of the vast majority of signals, the spinal cord can take care of some things on its own. These are known as 'spinal reflexes', and include responses like the knee-jerk reaction. They bypass the brain, which allows them to happen at super speed.

The peripheral nervous system is the network of nerves that feed the rest of your body, and it can be further divided into two parts: somatic and autonomic. The somatic nervous system looks after everything that you consciously feel and move, like clenching your leg muscles and sensing pain if you step on a nail. The autonomic system takes care of the things that go on in the background, like keeping your heart beating and your stomach churning.

Your nerve network
The nervous system sends electrical messages all over your body

Brain
The brainstem controls basic functions like breathing. The cerebellum coordinates movement, and the cerebrum is responsible for higher functions.

Thoracic nerves
There are 12 pairs of thoracic nerves, 11 of which lie between the ribs. They carry signals to the chest and abdomen.

Spinal cord
The spinal cord links the brain to the rest of the body, feeding messages backwards and forwards via branching nerves.

Ulnar nerve
These nerves run over the outside of the elbow, and are responsible for that odd 'funny bone' feeling.

Median nerve
This is one of the major nerves of the arm, and runs all the way down to the hand.

Lumbar nerves
There are five pairs of lumbar nerves, supplying the leg muscles.

Sacral nerves
There are five pairs of sacral nerves, supplying the ankles, as well as looking after bladder and bowel function.

Sciatic nerves
These are the longest spinal nerves in the body, with one running down each leg.

Oesophagus

o Sometimes known as the 'food pipe', this stretchy muscular tube links your mouth to your stomach. When you swallow, circular muscles contract to push food into your digestive tract, starting at the top and moving down in waves.

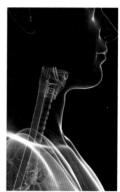

Pancreas

p This leaf-shaped organ plays two vital roles in digestion. It produces enzymes that break down food in the small intestine, and it makes the hormones insulin and glucagon, which regulate the levels of sugar in the blood.

© Thinkstock

Quadriceps

q There aren't many body parts that begin with the letter Q, but this bundle of four muscles in the upper leg is an important one. The quadriceps femoris connect the pelvis and thigh to the knee and shinbone, and are used to straighten the leg.

Ribcage

r This internal armour protects your heart and lungs, and performs a vital role in keeping your body supplied with oxygen. In total, the ribcage is made from 24 curved bones, which connect in pairs to the thoracic vertebrae of the spine at the back.

Seven of these pairs are called true ribs, and are linked at the front to a wide, flat bone called the sternum (or breastbone). The next three pairs, known as false ribs, connect to the sternum indirectly, and the final two don't link up at all, and are known as as floating ribs.

Not everyone has the same number of ribs, as sometimes the floating ribs are missing

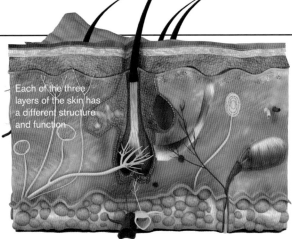

Each of the three layers of the skin has a different structure and function

Skin

s Your skin is the largest organ in your body. It is made up of three distinct layers: the epidermis on the outside, the dermis beneath, and the hypodermis right at the bottom.

The epidermis is waterproof, and is made up of overlapping layers of flattened cells. These are constantly being replaced by a layer of stem cells that sit just beneath. The epidermis also contains melanocytes, which produce the colour pigment melanin.

The dermis contains hair follicles, glands, nerves and blood vessels. It nourishes the top layer of skin, and produces sweat and sebum. Under this is a layer of supporting tissue called the hypodermis, which contains storage space for fat.

Tongue

Each papilla can have hundreds of taste buds, but some don't have any

Microvilli
Taste pore

Tongue

Papilla

t The tongue is a powerful muscle with several important functions. It is vital for chewing, swallowing, speech and even keeping your mouth clean, but its most well-known job is to taste.

The bumps on the tongue are not all taste buds; they are known as papillae, and there are four different types. At the very back of the tongue are the vallate papillae, each

Taste bud

containing around 250 taste buds. At the sides are the foliate papillae, with around 1,000 taste buds each. And at the tip are the fungiform (mushroom-shaped) papillae, with a whopping 1,600 taste buds each.

The rest of the bumps, covering most of the tongue, are known as filiform papillae, and do not have any taste buds at all.

Umbilical cord

u This spongy structure is packed with blood vessels, and connects a developing baby to its placenta. The placenta attaches to the wall of the mother's uterus, tapping into her blood supply to extract oxygen and nutrients. After birth, the cord dries up and falls away, leaving a scar called the belly button.

The umbilical cord is usually cut at birth, separating the baby from the placenta

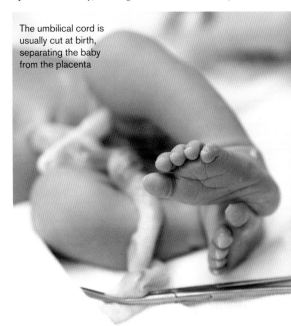

Vocal cords

V The vocal cords are folds of membrane found in the larynx, or voice box. They can be used to change the flow of air out of the lungs, allowing us to speak and sing. As air passes through the gap between the folds, they vibrate, producing sound.

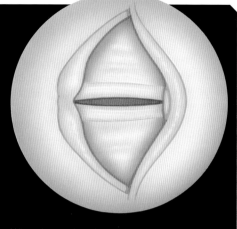

When the vocal cords are closed, pressure builds and they vibrate

Xiphoid process

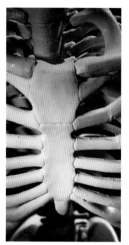

X This is the technical term used for the little lump that can be found at the bottom of your sternum, or breastbone. Medical professionals use the xiphoid process as a landmark in order to find the right place for chest compressions during CPR.

White blood cells

W These specialist cells make up your own personal army, tasked with defending your body from attack and disease. There are several different types, each with a unique role to play in keeping your body free of infection.

The first line of defence is called the innate immune system. These cells are the first ones on the scene, and they work to contain infections by swallowing and digesting bacteria, as well as killing cells that have been infected with viruses.

If the innate immune system can't keep the infection at bay, then they call in the second layer of defence – the adaptive immune system. These cells mount a stronger and more specific attack, and can even remember which pathogens they've fought before.

Yellow marrow

y There are two main types of bone marrow: yellow and red. Red marrow is responsible for producing new blood cells, while yellow marrow contains mainly fat. Red marrow gradually changes into yellow marrow as you get older.

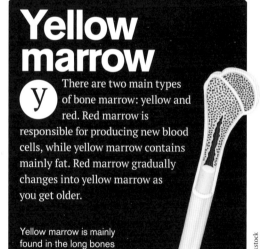

Yellow marrow is mainly found in the long bones of the arms and legs

© Thinkstock

Your immune army
Meet some of the cells that fight to keep you free from infection

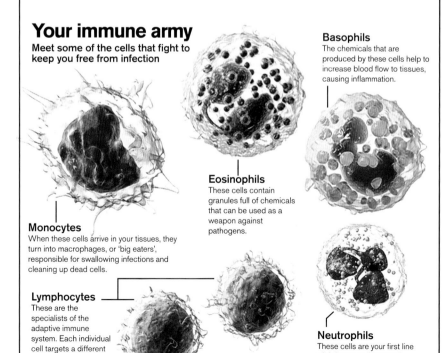

Basophils
The chemicals that are produced by these cells help to increase blood flow to tissues, causing inflammation.

Eosinophils
These cells contain granules full of chemicals that can be used as a weapon against pathogens.

Monocytes
When these cells arrive in your tissues, they turn into macrophages, or 'big eaters', responsible for swallowing infections and cleaning up dead cells.

Lymphocytes
These are the specialists of the adaptive immune system. Each individual cell targets a different enemy, delivering a deadly attack.

Neutrophils
These cells are your first line of defence against attack. They are present in large numbers in the blood.

Zygomaticus major

Z This is one of the key muscles responsible for your smile, joining the corner of the mouth to the cheekbone, and pulling your lips up and out. Depending on your anatomy, it is also the muscle responsible for cheek dimples.

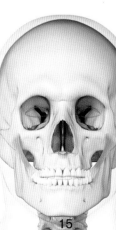

15

HUMAN ANATOMY

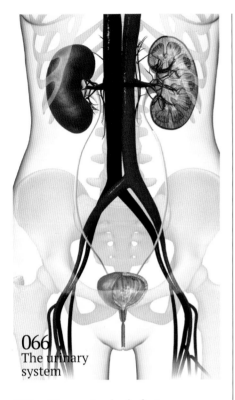

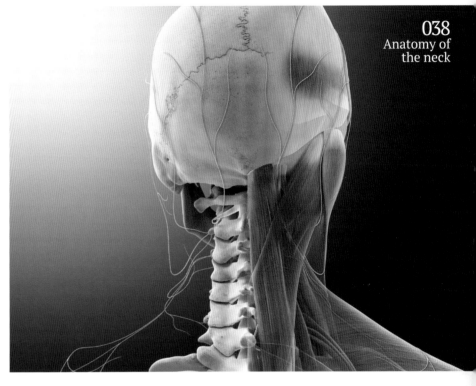

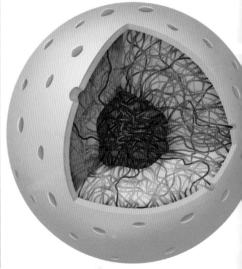

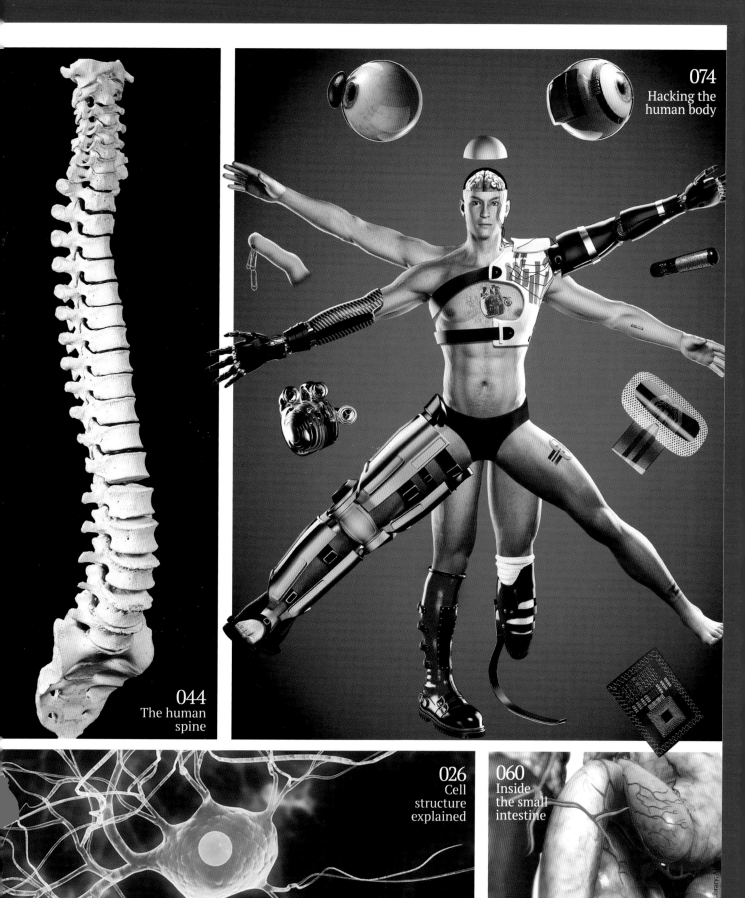

Hacking the
human body

044
The human
spine

026
Cell
structure
explained

060
Inside
the small
intestine

© Dreamstime; Science Photo Library;

50
Amazing facts about the human body

There are lots of medical questions everybody wants to ask but we just never get the chance... until now!

T he human body is the most complex organism we know and if humans tried to build one artificially, we'd fail abysmally. There's more we don't know about the body than we do know. This includes many of the quirks and seemingly useless traits that our species carry. However, not all of these traits are as bizarre as they may seem, and many have an evolutionary tale behind them.

Asking these questions is only natural but most of us are too embarrassed or never get the opportunity – so here's a chance to clear up all those niggling queries. We'll take a head-to-toe tour of the quirks of human biology, looking at everything from tongue rolling and why we are ticklish through to pulled muscles and why we dream.

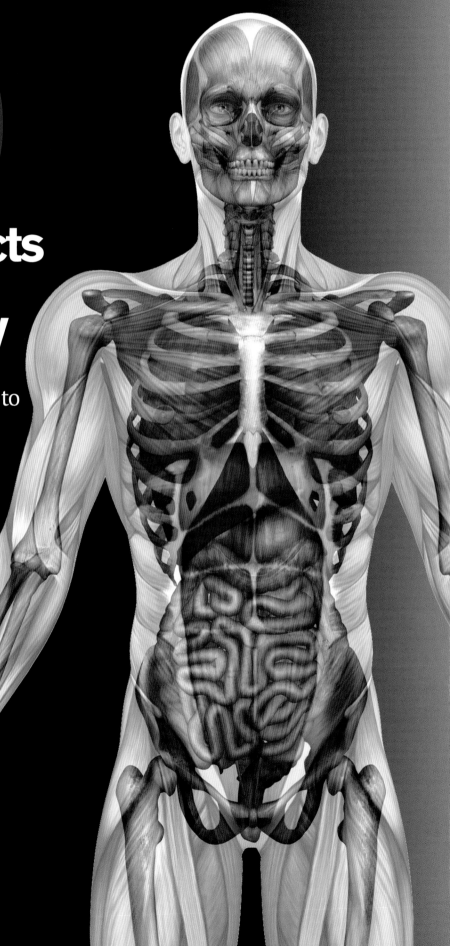

1 How do we think?

What are thoughts? This question will keep scientists, doctors and philosophers busy for decades to come. It all depends how you want to define the term 'thoughts'. Scientists may talk about synapse formation, pattern recognition and cerebral activation in response to a stimulus (seeing an apple and recognising it). Philosophers, and also many scientists, will argue that a network of neurons cannot possibly explain the many thousands of thoughts and emotions that we must deal with. A sports doctor might state that when you choose to run, you activate a series of well-trodden pathways that lead from your brain to your muscles in less than just a second.

There are some specifics we do know though – such as which areas of your brain are responsible for various types of thoughts and decisions.

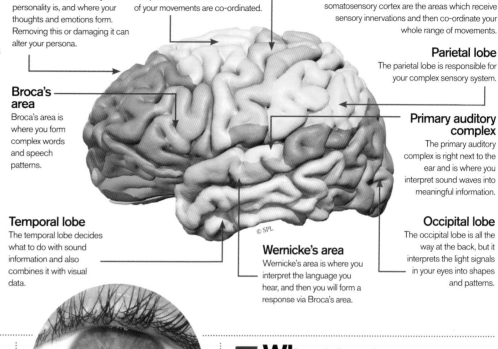

Frontal lobe
The frontal lobe is where your personality is, and where your thoughts and emotions form. Removing this or damaging it can alter your persona.

Pre-motor cortex
The pre-motor cortex is where some of your movements are co-ordinated.

Primary motor cortex
The primary motor cortex and the primary somatosensory cortex are the areas which receive sensory innervations and then co-ordinate your whole range of movements.

Parietal lobe
The parietal lobe is responsible for your complex sensory system.

Broca's area
Broca's area is where you form complex words and speech patterns.

Primary auditory complex
The primary auditory complex is right next to the ear and is where you interpret sound waves into meaningful information.

Temporal lobe
The temporal lobe decides what to do with sound information and also combines it with visual data.

Wernicke's area
Wernicke's area is where you interpret the language you hear, and then you will form a response via Broca's area.

Occipital lobe
The occipital lobe is all the way at the back, but it interprets the light signals in your eyes into shapes and patterns.

© SPL

2 In the mornings, do we wake up or open our eyes first?

Sleep is a gift from nature, which is more complex than you think. There are five stages of sleep which represent the increasing depths of sleep – when you're suddenly wide awake and your eyes spring open, it's often a natural awakening and you're coming out of rapid eye movement (REM) sleep; you may well remember your dreams. If you're coming out of a different phase, eg when your alarm clock goes off, it will take longer and you might not want to open your eyes straight away!

3 Do eyeballs grow like the rest of the body?

Only a small amount – this is actually why babies appear to be so beautiful, as their eyes are out of proportion and so appear bigger.

4 Why do we fiddle subconsciously? I'm constantly playing with my hair

This is a behavioural response – some people play with their hair when they're nervous or bored. For the vast majority of people such traits are perfectly normal. If they begin to interfere with your life, behavioural psychologists can help – but it's extremely rare that you'll end up there.

5 Why can some people roll their tongues but others can't?

Although we're often taught in school that tongue rolling is due to genes, the truth is likely to be more complex. There is likely to be an overlap of genetic factors and environmental influence. Studies on families and twins have shown that it simply cannot be a case of just genetic inheritance. Ask around – the fact that some people can learn to do it suggests that in at least some people it's environmental (ie a learned behaviour) rather than genetic (inborn).

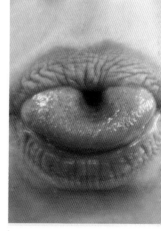

6 What is a pulse?

When you feel your own pulse, you're actually feeling the direct transmission of your heartbeat down your artery. You can only feel a pulse where you can compress an artery against a bone, eg the radial artery at the wrist. The carotid artery can be felt against the vertebral body, but beware, if press too hard and you can actually faint, press both at the same time and you'll cut off the blood to your brain and, as a protective mechanism, you'll definitely faint!

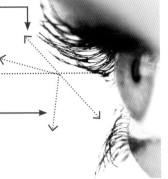

2D field
The areas from 120 to 180 degrees are seen as 2D as only one eye contributes, but we don't really notice.

3D field
The central 120-degree portion is the 3D part of our vision as both eyes contribute – this is the part we use the most.

© Matt Willman

7 What's my field of vision in degrees?

The human field of vision is just about 180 degrees. The central portion of this (approximately 120 degrees) is binocular or stereoscopic – ie both eyes contribute, allowing depth perception so that we can see in 3D. The peripheral edges are monocular, meaning that there is no overlap from the other eye so we see in 2D.

12 Why do we burp?

A burp is the bodies way of releasing gas naturally from your stomach. This gas has either been swallowed or is the result of something that you have ingested – such as a sparkling drink. The sound is vibrations which are taking place in the oesophageal sphincter, the narrowest part of the gastrointestinal tract.

8 What is the point of tonsils?

The tonsils are collections of lymphatic tissues which are thought to help fight off pathogens from the upper respiratory tract. However, the tonsils themselves can sometimes even become infected – leading to tonsillitis. The ones you can see at the back of your throat are just part of the ring of tonsils. You won't miss them if they're taken out for recurrent infections as the rest of your immune system will compensate.

© SPL

11 How fast does blood travel round the human body?

Your total 'circulating volume' is about five litres. Each red blood cell within this has to go from your heart, down the motorway-like arteries, through the back-road capillary system, and then back through the rush-hour veins to get back to your heart. The process typically takes about a minute. When you're in a rush and your heart rate shoots up, the time reduces as the blood diverts from the less-important structures (eg large bowel) to the more essential (eg muscles).

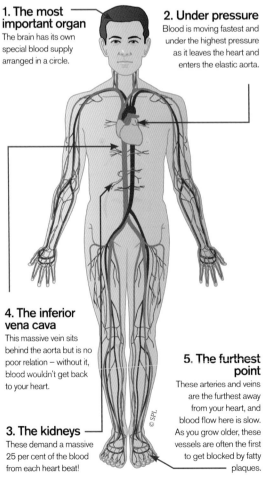

1. The most important organ
The brain has its own special blood supply arranged in a circle.

2. Under pressure
Blood is moving fastest and under the highest pressure as it leaves the heart and enters the elastic aorta.

4. The inferior vena cava
This massive vein sits behind the aorta but is no poor relation – without it, blood wouldn't get back to your heart.

5. The furthest point
These arteries and veins are the furthest away from your heart, and blood flow here is slow. As you grow older, these vessels are often the first to get blocked by fatty plaques.

3. The kidneys
These demand a massive 25 per cent of the blood from each heart beat!

© SPL

9 What are lips for?

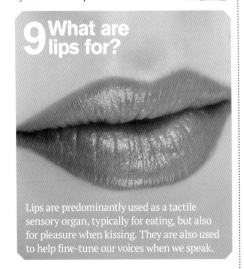

Lips are predominantly used as a tactile sensory organ, typically for eating, but also for pleasure when kissing. They are also used to help fine-tune our voices when we speak.

10 Why does it feel so weird when you hit your funny bone?

You're actually hitting the ulnar nerve as it wraps around the bony prominence of the 'humerus' bone, leading to a 'funny' sensation. Although not so funny as the brain interprets this sudden trauma as pain to your forearm and fingers!

13 How many inches of hair does the average person grow from their head each year?

It's different for everybody – your age, nutrition, health status, genes and gender all play a role. In terms of length, anywhere between 0.5-1 inch (1.2-2.5cm) a month might tends to be considered average,but don't be surprised if you're outside this range.

14 Why are everyone's fingerprints different?

Your fingerprints are fine ridges of skin in the tips of your fingers and toes. They are useful for improving the detection of small vibrations and to add friction for better grip. No two fingerprints are the same – either on your hands or between two people – and that's down to your unique set of genes.

15 Why do we only remember some dreams?

Dreams have fascinated humans for thousands of years. Some people think they are harmless while others think they are vital to our emotional wellbeing. Most people have four to eight dreams per night which are influenced by stress, anxiety and desires, but they remember very few of them. There is research to prove that if you awake from the rapid eye movement (REM) part of your sleep cycle, you're likely to remember your dreams more clearly.

16 Why, as we get older, does hair growth become so erratic?

Hair follicles in different parts of your body are actually programmed by your genes to do different things, eg the follicles on your arm produce hair much slower than those on your head. Men can go bald due to a combination of genes and hormonal changes, which may not happen in other areas (eg nasal hair).It's different for everybody!

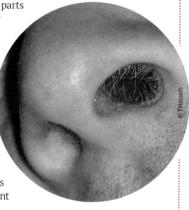

17 Why do we all have different coloured hair?

Most of it is down to the genes that result from when your parents come together to make you. Some hair colours win out (typically the dark ones) whereas some (eg blonde) are less strong in the genetic race.

18 Is it possible to keep your eyes open when you sneeze?

Your eyes remain shut as a defence mechanism to prevent the spray and nasal bacteria entering and infecting your eyes. The urban myth that your eyes will pop out if you keep them open is unlikely to happen – but keeping them shut will provide some protection against nasty bugs and viruses.

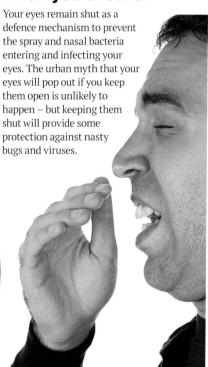

©Tristanb

19 What gives me my personality?

Researchers have spent their whole lives trying to answer this one. Your personality forms in the front lobes of your brain, and there are clear personality types. Most of it is your environment – that is, your upbringing, education, surroundings. However some of it is genetic, although it's unclear how much. The strongest research in this comes from studying twins – what influences one set of twins to grow up and be best friends, yet in another pair, one might become a professor and the other a murderer.

20 WHY DO MEN HAVE NIPPLES?

Men and women are built from the same template, and these are just a remnant of a man's early development.

21 WHAT'S THE POINT OF EYEBROWS?

Biologically, eyebrows can help to keep sweat and rainwater from falling into your eyes. More importantly in humans, they are key aids to non-verbal communication.

22 WHAT IS A BELLY BUTTON?

The umbilicus is where a baby's blood flows through to get to the placenta to exchange oxygen and nutrients with the mother's blood. Once out, the umbilical cord is clamped several centimetres away from the baby and left to fall off. No one quite knows why you'll get an 'innie' or an 'outie' – it's probably all just luck.

23 WHY IS IT THAT FINGERNAILS GROW MUCH FASTER THAN TOENAILS?

The longer the bone at the end of a digit, the faster the growth rate of the nail. However there are many other influences too – nutrition, sun exposure, activity, blood supply – and that's just to name a few.

24 WHY DOES MY ARM TINGLE AND FEEL HEAVY IF I FALL ASLEEP ON IT?

This happens because you're compressing a nerve as you're lying on your arm. There are several nerves supplying the skin of your arm and three supplying your hand (the radial, median and ulnar nerves), so depending on which part of your arm you lie on, you might tingle in your forearm, hand or fingers.

25 What makes some blood groups incompatible while others are universal?

Your blood type is determined by protein markers known as antigens on the surface of your red blood cells. You can have A antigens, B antigens, or none – in which case you're blood type O. However, if you don't have the antigen, your antibodies will attack foreign blood. If you're type A and you're given B, your antibodies attack the B antigens. However, if you're blood type AB, you can safely receive any type. Those who are blood group O have no antigens so can give blood to anyone, but they have antibodies to A and B so can only receive O back!

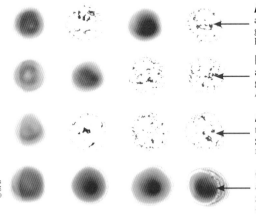

A You have A antigens and B antibodies. You can receive blood groups A and O, but can't receive B. You can donate to A and AB.

B You have B antigens and A antibodies. You can receive blood groups B and O, but can't receive A. You can donate to B and AB.

AB You have A and B antigens and no antibodies. You can receive blood groups A, B, AB and O (universal recipient), and can donate to AB.

O You have no antigens but have A and B antibodies. You can receive blood group O, but can't receive A, B or AB and can donate to all: A, B, AB and O.

26 What is a pulled muscle?

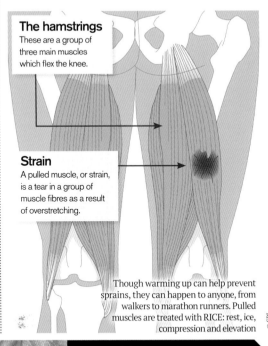

The hamstrings
These are a group of three main muscles which flex the knee.

Strain
A pulled muscle, or strain, is a tear in a group of muscle fibres as a result of overstretching.

Though warming up can help prevent sprains, they can happen to anyone, from walkers to marathon runners. Pulled muscles are treated with RICE: rest, ice, compression and elevation

28 What is the appendix? I've heard it has no use but can kill you…

The appendix is useful in cows for digesting grass and koala bears for digesting eucalyptus – koalas can have a 4m (13ft)-long appendix! In humans, however, the appendix has no useful function and is actually a remnant of our development. It typically measures 5-10cm (1.9-3.9in), but if it gets blocked it can get inflamed. If it isn't quickly removed, the appendix can burst and lead to widespread infection which can be lethal.

29 Why does people's skin turn yellow if they contract liver disease?

This yellow discolouration of the skin or the whites of the eyes is called jaundice. It is actually due to a buildup of bilirubin within your body, when normally this is excreted in the urine (hence why urine has a yellow tint). Diseases such as hepatitis and gallstones can lead to a buildup of bilirubin due to altered physiological processes, but there are other causes.

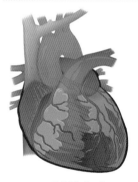

27 Which organ uses up the most oxygen?

The heart is the most efficient – it extracts 80 per cent of the oxygen from blood. But the liver gets the most blood – 40 per cent of the cardiac output compared to the kidneys, which get 25 per cent, and heart, which only receives 5 per cent.

30 What is the gag reflex?

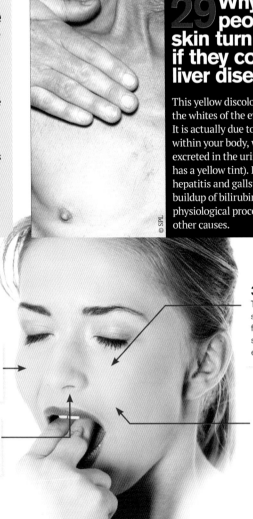

1. Foreign bodies
This is a protective mechanism to prevent food or foreign bodies entering the back of the throat at times other than swallowing.

2. Soft palate
The soft palate (the fleshy part of the mouth roof) is stimulated, sending signals down the glossopharyngeal nerve.

3. Vagus nerve
The vagus nerve is stimulated, leading to forceful contraction of the stomach and diaphragm to expel the object forwards.

4. The gag
This forceful expulsion leads to 'gagging', which can develop into retching and vomiting.

32 Why don't eyelashes keep growing?

Your eyelashes are formed from hair follicles, just like those on your head, arms and body. Each follicle is genetically programmed to function differently. Your eyelashes are programmed to grow to a certain length and even re-grow if they fall out, but they won't grow beyond a certain length, which is handy for seeing!

31 Why are we ticklish?

Light touches, by feathers, spiders, insects or other humans, can stimulate fine nerve-endings in the skin which send impulses to the somatosensory cortex in the brain. Certain areas are more ticklish – such as the feet – which may indicate that it is a defence mechanism against unexpected predators. It is the unexpected nature of this stimulus that means you can be tickled. Although you can give yourself goosebumps through light tickling, you can't make yourself laugh.

33 What makes us left-handed?

One side of the brain is more dominant over the other. Since each hemisphere of the brain controls the opposite side of your body, meaning the left controls the right side of your body. This is why right-handed people have stronger left brain hemispheres. However you can find an ambidextrous person, where hemispheres are co-dominant, and these people are equally capable with both right and left hands!

34 Could we survive on vitamins alone?

No, your body needs a diet balanced with vitamins, protein, minerals carbohydrates, and fat to survive. You can't cut one of these and expect your body to stay healthy. It is the proportions of these which keep us healthy and fit. You can get these from the five major food groups. Food charts can help with this balancing act.

35 Why do we get a high temperature when we're ill?

The immune response leads to inflammation and the release of inflammatory factors into your blood stream. These lead to an increased heart rate and blood flow, which increases your core body temperature – as if your body is doing exercise. This can lead to increased heat production and thus dehydration; for this reason, it's important to drink plenty of clear fluids when you're feeling unwell.

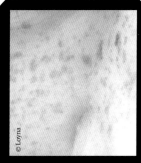

36 WHY DO SOME PEOPLE HAVE FRECKLES?

Freckles are concentrations of the dark skin pigment melanin in the skin. They typically occur on the face and shoulders, and are more common in light-skinned people. They are also a well-recognised genetic trait and become more dominant during sun-exposure.

37 WHAT IS A WART?

Warts are small, rough, round growths of the skin caused by the human papilloma virus. There are different types which can occur in different parts of the body, and they can be contagious. They commonly occur on the hands, but can also come up anywhere from the genitals to the feet!

38 WHY DO I TWITCH IN MY SLEEP?

This is known in the medical world as a myoclonic twitch. Although some researchers say these twitches are associated with stress or caffeine use, they are likely to be a natural part of the sleep process. If it happens to you, it's perfectly normal.

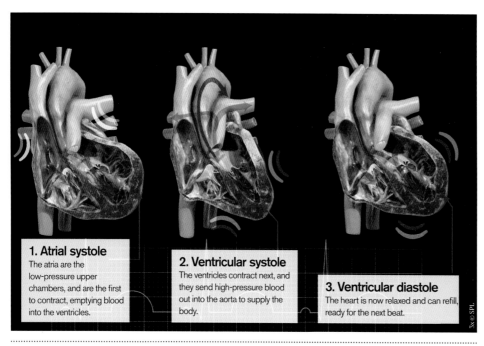

1. Atrial systole
The atria are the low-pressure upper chambers, and are the first to contract, emptying blood into the ventricles.

2. Ventricular systole
The ventricles contract next, and they send high-pressure blood out into the aorta to supply the body.

3. Ventricular diastole
The heart is now relaxed and can refill, ready for the next beat.

39 What triggers the heart and keeps it beating?

The heart keeps itself beating. The sinoatrial node (SAN) is in the wall of the right atrium of the heart, and is where the heartbeat starts. These beats occur due to changes in electrical currents as calcium, sodium and potassium move across membranes. The heart can beat at a rate of 60 beats per minute constantly if left alone. However – we often need it to go faster. The sympathetic nervous system sends rapid signals from the brain to stimulate the heart to beat faster when we need it to – in 'fight or flight' scenarios. If the SAN fails, a pacemaker can send artificial electrical signals to keep the heart going.

Definitions
Systole = contraction
Diastole = relaxation

40 Why do bruises go purple or yellow?

A bruise forms when capillaries under the skin leak and allow blood to settle in the surrounding tissues. The haemoglobin in red blood cells is broken down, and these by-products give a dark yellow, brown or purple discolouration depending on the volume of blood and colour of the overlying skin. Despite popular belief, you cannot age a bruise – different people's bruises change colour at different rates.

1. Damage to the blood vessels
After trauma such as a fall, the small capillaries are torn and burst.

3. Discolouration
Haemoglobin is then broken down into its smaller components, which are what give the dark discolouration of a bruise.

2. Blood leaks into the skin
Blood settles into the tissues surrounding the vessel. The pressure from the bruise then helps stem the bleeding.

41 Why does cutting onions make us cry?

Onions make your eyes water due to their expulsion of an irritant gas once cut. This occurs as when an onion is cut with a knife, many of its internal cells are broken down, allowing enzymes to break down amino acid sulphoxides and generate sulphenic acids. These sulphenic acids are then rearranged by another enzyme and, as a direct consequence, syn-propanethial-S-oxide gas is produced, which is volatile. This volatile gas then diffuses in the air surrounding the onion, eventually reaching the eyes of the cutter, where it proceeds to activate sensory neurons and create a stinging sensation. As such, the eyes then follow protocol and generate tears from their tear glands in order to dilute and remove the irritant. Interestingly, the volatile gas generated by cutting onions can be largely mitigated by submerging the onion in water prior to or midway through cutting, with the liquid absorbing much of the irritant.

Inli Masriera

42 What is the little triangle shape on the side of the ear?

This is the tragus. It serves no major function that we know of, but it may help to reflect sounds into the ear to improve hearing.

© David Benbennick

43 When we're tired, why do we get bags under our eyes?

Blood doesn't circulate around your body as efficiently when you're asleep so excess water can pool under the eyes, making them puffy. Fatigue, nutrition, age and genes also cause bags.

44 Why do more men go bald than women?

'Simple' male pattern baldness is due to a combination of genetic factors and hormones. The most implicated hormone is testosterone, which men have high levels of but women have low levels of, so they win (or lose?) in this particular hormone contest!

45 Why do we blink?

Blinking helps keep your eyes clean and moist. Blinking spreads secretions from the tear glands (lacrimal fluids) over the surface of the eyeball, keeping it moist and also sweeping away small particles such as dust.

47 Why do we get itchy?

Itching is caused by the release of a transmitter called histamine from mast cells which circulate in your body. These cells are often released in response to a stimulus, such as a bee sting or an allergic reaction. They lead to inflammation and swelling, and send impulses to the brain via nerves which causes the desire to itch.

48 Why do some hereditary conditions skip a generation?

Genes work in pairs. Some genes are 'recessive' and if paired with a 'dominant' half, they won't shine through. However, if two recessive genes combine (one from your mother and one from your father), the recessive trait will show through.

49 Why do amputees sometimes feel pain in their amputated limbs?

This is 'phantom limb pain' and can range from a mild annoyance to a debilitating pain. The brain can sometimes struggle to adjust to the loss of a limb, and it can still 'interpret' the limb as being there. Since the nerves have been cut, it interprets these new signals as pain. There isn't a surgical cure as yet, though time and special medications can help lessen the pain.

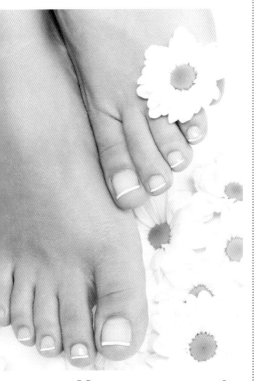

46 How come most people have one foot larger than the other?

Most people's feet are different sizes – in fact the two halves of most people's bodies are different! We all start from one cell, but as the cells multiply, genes give them varying characteristics.

50 Which muscle produces the most powerful contraction relative to its size?

The gluteus maximus is the largest muscle and forms the bulk of your buttock. The heart (cardiac muscle) is the hardest-working muscle, as it is constantly beating and clearly can never take a break! However the strongest muscle based on weight is the masseter. This is the muscle that clenches the jaw shut – put a finger over the lowest, outer part of your jaw and clench your teeth and you'll feel it.

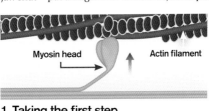

1. Taking the first step
Muscle contraction starts with an impulse received from the nerves supplying the muscle – an action potential. This action potential causes calcium ions to flood across the protein muscle fibres. The muscle fibres are formed from two key proteins: actin and myosin.

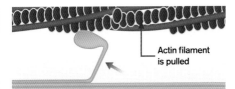

2. Preparation
The calcium binds to troponin which is a receptor on the actin protein. This binding changes the shape of tropomyosin, another protein which is bound to actin. These shape changes lead to the opening of a series of binding sites on the actin protein.

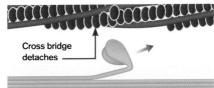

3. Binding
Now the binding sites are free on actin, the myosin heads forge strong bonds in these points. This leads to the contraction of the newly formed protein complex; when all of the proteins contract, the muscle bulk contracts.

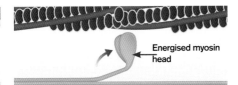

4. Unbinding
When the energy runs out, the proteins lose their strong bonds and disengage, and from there they return to their original resting state. This is the unbinding stage.

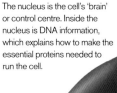

Cell structure explained

The human body has over 75 trillion cells, but what are they and how do they work?

Cells are life and cells are alive. You are here because every cell inside your body has a specific function and a very specialised job to do. There are many different types of cell, each one working to keep the body's various systems operating. A single cell is the smallest unit of living material in the body capable of life. When grouped together in layers or clusters, however, cells with similar jobs to do form tissue, such as skin or muscle. To keep these cells working, there are thousands of chemical reactions going on all the time.

All animal cells contain a nucleus, which acts like a control hub telling the cell what to do and contains the cell's genetic information (DNA). Most of the material within a cell is a watery, jelly-like substance called cytoplasm (cyto means cell), which circulates around the cell and is held in by a thin external membrane, which consists of two layers. Within the cytoplasm is a variety of structures called organelles, which all have different tasks, such as manufacturing proteins – the cell's key chemicals. One vital example of an organelle is a ribosome; these numerous structures can be found either floating around in the cytoplasm or attached to internal membranes. Ribosomes are crucial in the production of proteins from amino acids.

In turn, proteins are essential to building your cells and carrying out the biochemical reactions the body needs in order to grow and develop and also to repair itself and heal.

Cell membrane
Surrounding and supporting each cell is a plasma membrane that controls everything that enters and exits.

Nucleus
The nucleus is the cell's 'brain' or control centre. Inside the nucleus is DNA information, which explains how to make the essential proteins needed to run the cell.

Ribosomes
These tiny structures make proteins and can be found either floating in the cytoplasm or attached like studs to the endoplasmic reticulum, which is a conveyor belt-like membrane that transports proteins around the cell.

Endoplasmic reticulum
The groups of folded membranes (canals) connecting the nucleus to the cytoplasm are called the endoplasmic reticulum (ER). If studded with ribosomes the ER is referred to as 'rough' ER; if not it is known as 'smooth' ER. Both help transport materials around the cell but also have differing functions.

Smooth endoplasmic reticulum

Rough endoplasmic reticulum (studded with ribosomes)

Mitochondria
These organelles supply cells with the energy necessary for them to carry out their functions. The amount of energy used by a cell is measured in molecules of adenosine triphosphate (ATP). Mitochondria use the products of glucose metabolism as fuel to produce the ATP.

Golgi body
Another organelle, the Golgi body is one that processes and packages proteins, including hormones and enzymes, for transportation either in and around the cell or out towards the membrane for secretion outside the cell where it can enter the bloodstream.

Cell anatomy

Cytoplasm
This is the jelly-like substance – made of water, amino acids and enzymes – found inside the cell membrane. Within the cytoplasm are organelles such as the nucleus, mitochondria and ribosomes, each of which performs a specific role, causing chemical reactions in the cytoplasm.

Pore

© Science Photo Library

Lysosomes
This digestive enzyme breaks down unwanted substances and worn-out organelles that could harm the cell by digesting the product and then ejecting it outside the cell.

Types of human cell

So far around 200 different varieties of cell have been identified, and they all have a very specific function to perform. Discover the main types and what they do…

NERVE CELLS
The cells that make up the nervous system and the brain are nerve cells or neurons. Electrical messages pass between nerve cells along long filaments called axons. To cross the gaps between nerve cells (the synapse) that electrical signal is converted into a chemical signal. These cells enable us to feel sensations, such as pain, and they also enable us to move.

BONE CELLS
The cells that make up bone matrix – the hard structure that makes bones strong – consist of three main types. Your bone mass is constantly changing and reforming and each of the three bone cells plays its part in this process. First the osteoblasts, which come from bone marrow, build up bone mass and structure. These cells then become buried in the matrix at which point they become known as osteocytes. Osteocytes make up around 90 per cent of the cells in your skeleton and are responsible for maintaining the bone material. Finally, while the osteoblasts add to bone mass, osteoclasts are the cells capable of dissolving bone and changing its mass.

PHOTORECEPTOR CELLS
The cones and rods on the retina at the back of the eye are known as photoreceptor cells. These contain light-sensitive pigments that convert the image that enters the eye into nerve signals, which the brain interprets as pictures. The rods enable you to perceive light, dark and movement, while the cones bring colour to your world.

LIVER CELLS
The cells in your liver are responsible for regulating the composition of your blood. These cells filter out toxins as well as controlling fat, sugar and amino acid levels. Around 80 per cent of the liver's mass consists of hepatocytes, which are the liver's specialised cells that are involved with the production of proteins and bile.

MUSCLE CELLS
There are three types of muscle cell – skeletal, cardiac and smooth – and each differs depending on the function it performs and its location in the body. Skeletal muscles contain long fibres that attach to bone. When triggered by a nerve signal, the muscle contracts and pulls the bone with it, making you move. We can control skeletal muscles because they are voluntary. Cardiac muscles,

meanwhile, are involuntary, which is fortunate because they are used to keep your heart beating. Found in the walls of the heart, these muscles create their own stimuli to contract without input from the brain. Smooth muscles, which are pretty slow and also involuntary, make up the linings of hollow structures such as blood vessels and your digestive tract. Their wave-like contraction aids the transport of blood around the entire body and the digestion of food.

FAT CELLS
These cells – also known as adipocytes or lipocytes – make up your adipose tissue, or body fat, which can cushion, insulate and protect the body. This tissue is found beneath your skin and also surrounding your other organs. The size of a fat cell can increase or decrease depending on the amount of energy it stores. If we gain weight the cells fill with more watery fat, and eventually the number of fat cells will begin to increase. There are two types of adipose tissue: white and brown. The white adipose tissue stores energy and insulates the body by maintaining body heat. The brown adipose tissue, on the other hand, can actually create heat and isn't burned for energy – this is why animals are able to hibernate for months on end without food.

EPITHELIAL CELLS
Epithelial cells make up the epithelial tissue that lines and protects your organs and constitute the primary material of your skin. These tissues form a barrier between the precious organs and unwanted pathogens or other fluids. As well as covering your skin, you'll find epithelial cells inside your nose, around your lungs and in your mouth.

RED BLOOD CELLS
Unlike all the other cells in your body, your red blood cells (also known as erythrocytes) do not contain a nucleus. You are topped up with around 25 trillion red blood cells – that's a third of all your cells, making them the most common cell found in your body. Formed in the bone marrow, these cells are important because they carry oxygen to all the different tissues in your body. Oxygen is carried in haemoglobin, a pigmented protein that gives the blood cells their recognisable red colour.

Inside a nucleus

Dissecting the control centre of a cell

Surrounded by cytoplasm, the nucleus contains a cell's DNA and controls all of its functions and processes such as movement and reproduction.

There are two main types of cell: eukaryotic and prokaryotic. Eukaryotic cells contain a nucleus while prokaryotic do not. Some eukaryotic cells have more than one nucleus – called multinucleate cells – occurring when fusion or division creates two or more nuclei.

At the heart of a nucleus you'll find the nucleolus; this particular area is essential in the formation of ribosomes.

Ribosomes are responsible for making proteins out of amino acids which take care of growth and repair.

The nucleus is the most protected part of the cell. In animal cells it is located near its centre and away from the membrane for maximum cushioning. As well as the jelly-like cytoplasm around it, the nucleus is filled with nucleoplasm, a viscous liquid which maintains its structural integrity.

Conversely, in plant cells, the nucleus is more sporadically placed. This is due to the fact that a plant cell has a larger vacuole and there is added protection which is granted by a cell wall.

Take a peek at what's happening inside the 'brain' of a eukaryotic cell

❶ Nuclear pore
These channels control the movement of molecules between the nucleus and cytoplasm.

❷ Nuclear envelope
Acts as a wall to protect the DNA within the nucleus and regulates cytoplasm access.

❸ Nucleolus
Made up of protein and RNA, this is the heart of the nucleus which manufactures ribosomes.

❹ Nucleoplasm
This semi-liquid, semi-jelly material surrounds the nucleolus and keeps the organelle's structure.

❺ Chromatin
Produces chromosomes and aids cell division by condensing DNA molecules.

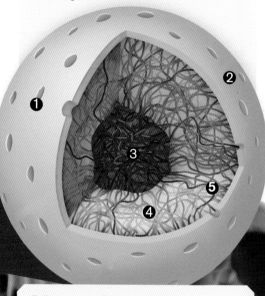

Nucleus in context

Explore the larger body that a nucleus rules over and meet its 'cellmates'

Nucleus

Ribosomes
Made up of two separate entities, ribosomes make proteins to be used both inside and outside the cell.

Golgi apparatus
Named after the Italian biologist Camillo Golgi, they create lysosomes and also organise the proteins for secretion.

Lysosome
Small and spherical, this organelle contains digestive enzymes that attack invading bacteria.

Mitochondrion
Double membraned, this produces energy for the cell by breaking down nutrients via cellular respiration.

How do cells survive without a nucleus?

Prokaryotic cells are actually much more basic than their eukaryotic counterparts. Not only are they up to 100 times smaller but they also are mainly a comprising species of bacteria, prokaryotic cells have fewer functions than other cells, so they do not require a nucleus to act as the control centre for the organism.

Instead, these cells have their DNA moving around the cell rather than being housed in a nucleus. They have no chloroplasts, no membrane-bound organelles and they don't undertake cell division in the form of mitosis or meiosis like eukaryotic cells do.

Prokaryotic cells divide asexually with DNA molecules replicating themselves in a process that is known as binary fission.

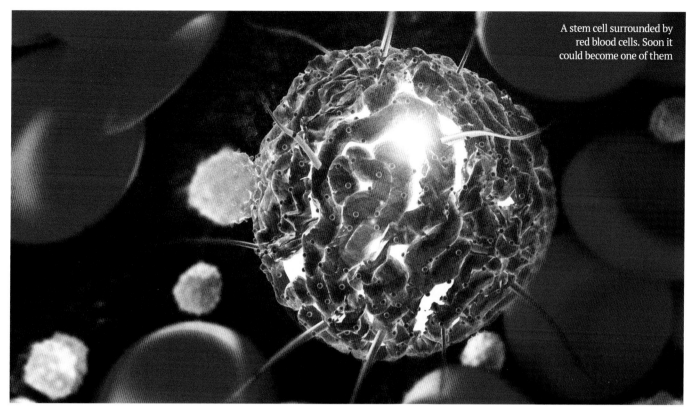

A stem cell surrounded by red blood cells. Soon it could become one of them

What are stem cells?

Understand how these building blocks bring new life

Stem cells are incredibly special because they have the potential to become any kind of cell in the body, from red blood cells to brain cells. They are essential to life and growth, as they repair tissues and replace dead cells. Skin, for example, is constantly replenished by skin stem cells.

Stem cells begin their life cycle as generic, featureless cells that don't contain tissue-specific structures, such as the ability to carry oxygen. Stem cells become specialised through a process called differentiation. This is triggered by signals inside and outside the cell. Internal signals come from strands of DNA that carry information for all cellular structures, while external signals include chemicals from nearby cells. Stem cells can replicate many times – known as proliferation – while others such as nerve cells don't divide at all.

There are two stem cell types, as Professor Paul Fairchild, co-director of the Oxford Stem Cell Institute at Oxford Martin School explains: "Adult stem cells are multipotent, which means they are able to produce numerous cells that are loosely related, such as stem cells in the bone marrow can generate cells that make up the blood," he says. "In contrast, pluripotent stem cells, found within developing embryos, are able to make any one of the estimated 210 cell types that make up the human body."

This fascinating ability to transform and divide has made stem cells a rich source for medical research. Once their true potential has been harnessed, they could be used to treat a huge range of diseases and disabilities.

Cloning cells

Scientists can reprogram cells to forget their current role and become pluripotent cells indistinguishable from early embryonic stem cells. Induced pluripotent stem cells (IPSCs) can be used to take on the characteristics of nearby cells.

IPSCs are more reliable than stem cells grown from a donated embryo because the body is more likely to accept self-generated cells. IPSCs can treat degenerative conditions such as Parkinson's disease and baldness, which are caused by cells dying without being replaced. The IPSCs fill those gaps in order to restore the body's systems.

Professor Fairchild explains the process to us: "By deriving these cells from individuals with rare conditions, we are able to model the condition in the laboratory and investigate the effects of new drugs on that disease."

Research on cloning cells can help cure diseases

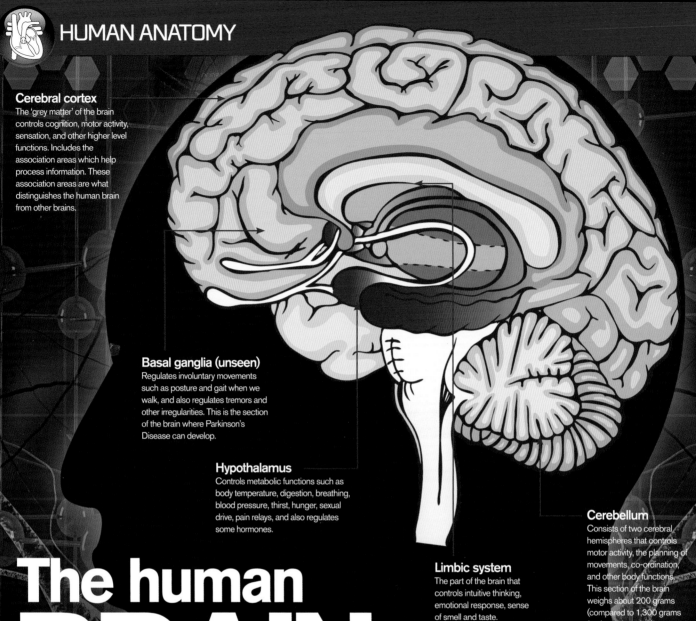

Cerebral cortex
The 'grey matter' of the brain controls cognition, motor activity, sensation, and other higher level functions. Includes the association areas which help process information. These association areas are what distinguishes the human brain from other brains.

Basal ganglia (unseen)
Regulates involuntary movements such as posture and gait when we walk, and also regulates tremors and other irregularities. This is the section of the brain where Parkinson's Disease can develop.

Hypothalamus
Controls metabolic functions such as body temperature, digestion, breathing, blood pressure, thirst, hunger, sexual drive, pain relays, and also regulates some hormones.

Cerebellum
Consists of two cerebral hemispheres that controls motor activity, the planning of movements, co-ordination, and other body functions. This section of the brain weighs about 200 grams (compared to 1,300 grams for the main cortex).

Limbic system
The part of the brain that controls intuitive thinking, emotional response, sense of smell and taste.

The human
BRAIN

Described as the most complex thing in the universe, our brains are truly astonishing

The brain makes up just two per cent of our total body weight, but crammed inside are approximately 86 billion neurons, surrounded by 180,000 kilometres of insulated fibres connected at 100 trillion synapses. It's a vast biological supercomputer.

The cells in the brain communicate using electrical signals. When a message is sent, thousands of microscopic channels open, allowing positively charged ions to flood across the membrane. Afterwards, more than 1 million miniature pumps in each cell move the ions back again ready for the next impulse.

The cell bodies of the neurons, and their connections, are contained within the grey matter, which consumes 94 per cent of the oxygen delivered to the brain. Different areas are responsible for different functions, and wiring them together is a fatty network of fibres called white matter.

When a signal reaches the end of a nerve cell, tiny packets of chemical signals spill out onto the surrounding neurons. These connections, called synapses, allow messages to be passed from one cell to the next. Each neuron can receive thousands of inputs, coordinating them in time and space, and by type of chemical, to decide what to do next.

Scientists have been electrically and chemically stimulating the brain to see how it responds to different signals, recording electrical activity to map thoughts and using imaging like functional MRI to track the blood flow increases that reveal when nerve cells are firing. The cells of the brain can also be studied inside the lab. Thanks to these investigations we know more about this incredible structure than ever before, but our understanding is only just beginning. There is so much more to learn.

Brain development

From a single cell to an incredibly intricate network in just nine months

Within weeks of fertilisation, neural progenitors start to form; these stem cells will go on to become all of the cells of the central nervous system. They organise into a neural tube when the embryo is barely the size of a pen tip, and then patterning begins, laying out the structural organisation of the brain and spinal cord. At its peak growth rate, the developing brain can generate 250,000 new neurons every minute. By the time a baby is born, the process still isn't complete. But, by the age of two, the brain is 80 per cent of its adult size.

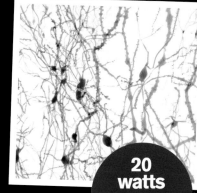

20 watts
Your brain is incredibly efficient, using less energy than a standard lightbulb.

Pyramidal neurons, like these, are found in the hippocampus, cortex and amygdala

Brain formation

This astonishing structure is formed and refined as pregnancy progresses

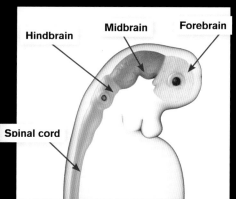

4 weeks
Brain development starts just three weeks after fertilisation. The first structure is the neural tube, which divides into regions that later become the forebrain, midbrain, hindbrain and spinal cord.

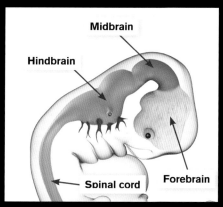

6 weeks
The pattern of the brain and spinal cord is now laid out and is gradually refined, controlled by gradients of signalling molecules that assign different areas for different functions.

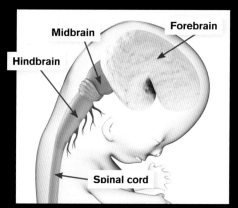

11 weeks
As the embryo becomes larger, the brain continues to increase in size and neurons migrate and organise. The surface of the brain gradually begins to fold. At this point, a foetus only measures about five centimetres in length.

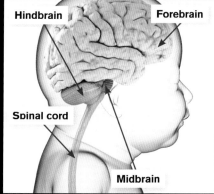

Birth
Before a baby is born, around half of the nerve cells in the brain are lost and connections are pruned, leaving only the most useful. This process continues after birth.

Why the brain is wrinkled

The brain folds in on itself to cram in more processing power

The folds and pockets of our brains are a biological rarity that we only share with a few other species, including dolphins, some primates and elephants. It's a clever evolutionary adaptation that allows intelligent species to squash a huge amount of cortical tissue into a small space, allowing enormous brainpower to be crammed into our relatively small skulls.

Folding starts during the second trimester of pregnancy, creating ridges (gyri) and fissures (sulci), but the biology behind the distinctive wrinkles is stranger than you might think. The organisation of the brain is determined by complex cascades of chemical signals, but the overall shape seems to be the result of simple physics. Grey matter sits on the outside of the brain and, during development, its growth rapidly outpaces the growth of white matter underneath. This puts mechanical stress on the structure, forcing the outside to buckle and curl.

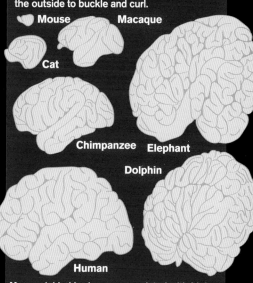

More wrinkled brains are associated with higher intelligence (brain sizes not to scale)

"Our brains contain 86 billion neurons and 180,000 kilometres of fibres"

The brain can regenerate
Research has shown that certain areas of the adult brain can continue to produce new neurons, a process known as neurogenesis.

© Thinkstock; Illustrations by Jo Smolaga

Making memories

The brain can store around 1 million gigabytes of data

A team at the Salk Institute in California estimate that the brain can store around 1 petabyte of information, stuffed into the connections between nerve cells. That's around 2,000 years worth of MP3 music or 223,000 DVDs. And, incredibly, it's possible to watch memories being made.

The Weizmann Institute in Israel and UCLA in the US captured memory formation in action. Patients watched clips of videos and were then asked to recall what they'd seen. The neurons that lit up when they watched the first time lit up again as they relived the experience inside of their heads – a bit like an echo in the brain.

Recent research from the US and Japan suggests that these echoes are actually stored twice – once in the hippocampus and again in the cortex. The hippocampus handles short-term storage and gradually forgets, but as it does so it helps to reinforce the memory in the cortex, allowing long-term recall.

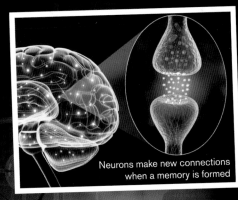

Neurons make new connections when a memory is formed

Self-cleaning brains

We have a built-in system to clear toxic waste from between our brain cells

Sleep is one of the brain's great mysteries, but research on mice has revealed an intriguing night time cleanup system. The brain is shielded by a barrier made and maintained by cells called astrocytes. They hug the blood vessels, controlling what's allowed in and out, and a space between the vessel wall and these cells seems to play a crucial role in keeping the brain clean.

At night, the astrocytes relax their grip and the space fills up with a clear liquid called cerebrospinal fluid (CSF). It's pushed along by the movement of the blood vessels underneath, swishing up through the astrocytes and out into the spaces between brain cells. As it passes, it picks up waste and debris, carrying the particles back towards the bloodstream so that they can be removed from the brain.

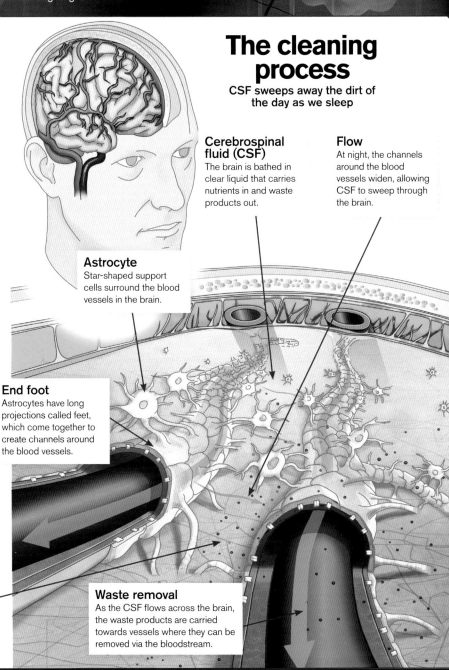

The cleaning process

CSF sweeps away the dirt of the day as we sleep

Cerebrospinal fluid (CSF)
The brain is bathed in clear liquid that carries nutrients in and waste products out.

Flow
At night, the channels around the blood vessels widen, allowing CSF to sweep through the brain.

Astrocyte
Star-shaped support cells surround the blood vessels in the brain.

End foot
Astrocytes have long projections called feet, which come together to create channels around the blood vessels.

Waste
Brain cells are constantly creating waste products that can cause damage if they're allowed to build up.

Waste removal
As the CSF flows across the brain, the waste products are carried towards vessels where they can be removed via the bloodstream.

How do nerves work?

Nerves carry signals throughout the body – a chemical superhighway

Nerves are the transmission cables that carry brain waves in the human body, says Sol Diamond, an assistant professor at the Thayer School of Engineering at Dartmouth. According to Diamond, nerves communicate these signals from one point to another, whether from your toenail up to your brain or from the side of your head.

Nerve transmissions
Some nerve transmissions travel great distances through the human body, others travel short distances – both use a de-polarisation to create the circuit. De-polarisation is like a wound-up spring that releases stored energy once it is triggered.

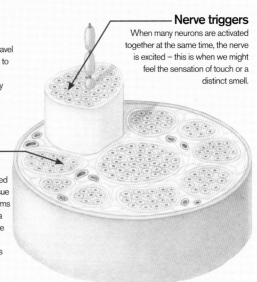

Nerve triggers
When many neurons are activated together at the same time, the nerve is excited – this is when we might feel the sensation of touch or a distinct smell.

Myelinated and un-mylinated
Some nerves are myelinated (or insulated) with fatty tissue that appears white and forms a slower connection over a longer distance. Others are un-myelinated and are un-insulated. These nerves travel shorter distances.

© DK Images

What does the spinal cord do?

The spinal cord actually is part of the brain and plays a major role

Scientists have known for the past 100 years or so that the spinal cord is actually part of the brain. According to Melillo, while the brain has grey matter on the outside (protected by the skull) and protected white matter on the inside, the spinal cord is the reverse: the grey matter is inside the spinal cord and the white matter is outside.

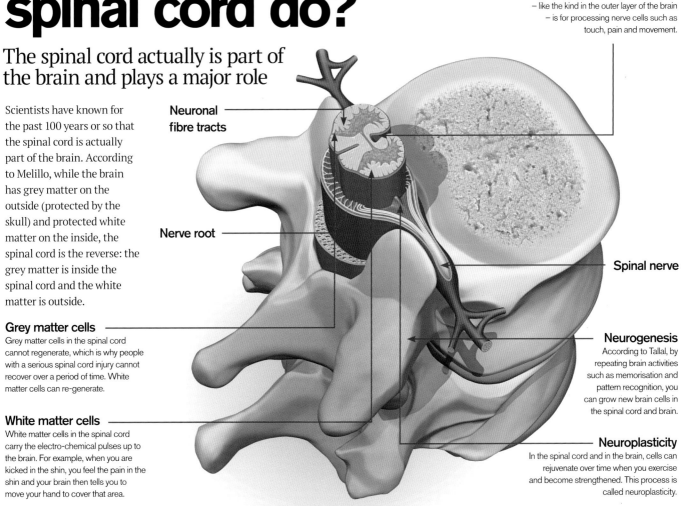

Spinal cord core
In the core of the spinal cord, grey matter – like the kind in the outer layer of the brain – is for processing nerve cells such as touch, pain and movement.

Neuronal fibre tracts

Nerve root

Spinal nerve

Neurogenesis
According to Tallal, by repeating brain activities such as memorisation and pattern recognition, you can grow new brain cells in the spinal cord and brain.

Neuroplasticity
In the spinal cord and in the brain, cells can rejuvenate over time when you exercise and become strengthened. This process is called neuroplasticity.

Grey matter cells
Grey matter cells in the spinal cord cannot regenerate, which is why people with a serious spinal cord injury cannot recover over a period of time. White matter cells can re-generate.

White matter cells
White matter cells in the spinal cord carry the electro-chemical pulses up to the brain. For example, when you are kicked in the shin, you feel the pain in the shin and your brain then tells you to move your hand to cover that area.

Inside the human eye

Uncovering one of the most complex constructs in the natural world

The structure of the human eye is so incredibly complex that it's actually hard to believe that it's not the product of intelligent design. But by looking at and studying the eyes of various other animals, scientists have been able to show that eyes have evolved very gradually from just a simple light-dark sensor over the course of around 100 million years. The eye functions in a very similar way to a camera, with an opening through which the light enters, a lens for focusing and a light-sensitive membrane.

The amount of light that enters the eye is controlled by the circular and radial muscles in the iris, which contract and relax to alter the size of the pupil. The light first passes through a tough protective sheet called the cornea, and then moves into the lens. This adjustable structure bends the light, focusing it down to a point on the retina, at the back of the eye.

The retina is covered in millions of light-sensitive receptors known as rods and cones. Each receptor contains pigment molecules, which change shape when they are hit by light, which triggers an electrical message that then travels to the brain via the optic nerve.

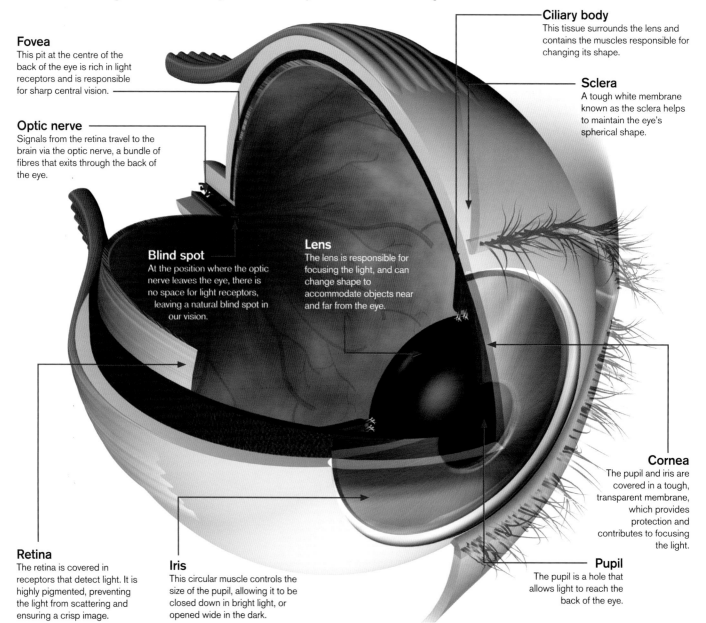

Ciliary body
This tissue surrounds the lens and contains the muscles responsible for changing its shape.

Sclera
A tough white membrane known as the sclera helps to maintain the eye's spherical shape.

Fovea
This pit at the centre of the back of the eye is rich in light receptors and is responsible for sharp central vision.

Optic nerve
Signals from the retina travel to the brain via the optic nerve, a bundle of fibres that exits through the back of the eye.

Blind spot
At the position where the optic nerve leaves the eye, there is no space for light receptors, leaving a natural blind spot in our vision.

Lens
The lens is responsible for focusing the light, and can change shape to accommodate objects near and far from the eye.

Cornea
The pupil and iris are covered in a tough, transparent membrane, which provides protection and contributes to focusing the light.

Retina
The retina is covered in receptors that detect light. It is highly pigmented, preventing the light from scattering and ensuring a crisp image.

Iris
This circular muscle controls the size of the pupil, allowing it to be closed down in bright light, or opened wide in the dark.

Pupil
The pupil is a hole that allows light to reach the back of the eye.

How the eye focuses

The tiny rings of muscle that make your vision sharp

Cameras and human eyes both focus light using a lens. This structure bends the incoming wavelengths so that they hit the right spot on a photographic plate, or on the back of the eye. A camera lens is made from solid glass, and focuses on near and distant objects by physically moving closer or further away. A biological lens is actually squishy, and it focuses by physically changing shape.

In the eye, this process is known as 'accommodation', and is controlled by a ring of smooth muscle called the ciliary muscle. This is attached to the lens by fibres known as suspensory ligaments. When the muscle is relaxed, the ligaments pull tight, stretching the lens until it is flat and thin. This is perfect for looking at objects in the distance.

When the ciliary muscle contracts, the ligaments loosen, allowing the lens to become fat and round. This is better for looking at objects that are nearby. The coloured part of the eye (called the iris) controls the size of the pupil and ensures the right amount of light gets through the lens.

Lens
The lens is responsible for focusing the light on the back of the eye.

Accommodation explained
How the lens changes its shape to focus on near and distant objects

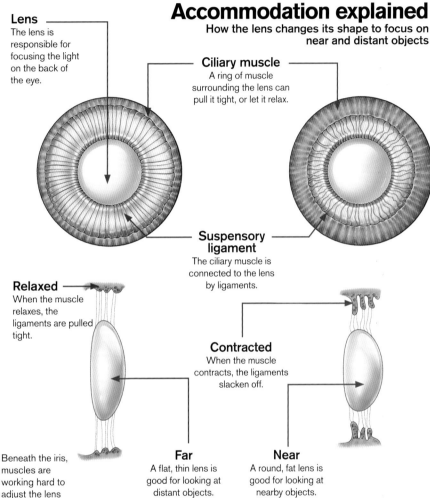

Ciliary muscle
A ring of muscle surrounding the lens can pull it tight, or let it relax.

Suspensory ligament
The ciliary muscle is connected to the lens by ligaments.

Relaxed
When the muscle relaxes, the ligaments are pulled tight.

Contracted
When the muscle contracts, the ligaments slacken off.

Beneath the iris, muscles are working hard to adjust the lens

Far
A flat, thin lens is good for looking at distant objects.

Near
A round, fat lens is good for looking at nearby objects.

Seeing in three dimensions

Each eye sees a slightly different image, allowing the brain to perceive depth

Our eyes are only able to produce two-dimensional images, but with some clever internal processing, the brain is able to build these flat pictures into a three-dimensional view. Our eyes are positioned about five centimetres (two inches) apart, so each sees the world from a slightly different angle. The brain then compares the two pictures, using the differences to create the illusion of depth.

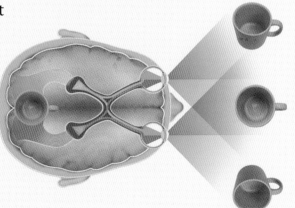

Individual image
Due to the positioning of our eyes, when objects are closer than about 5.5m (18ft) away, each eye sees a slightly different angle.

Combined image
The incoming signals from both eyes are compared in the brain, and the subtle differences are used to create a three-dimensional image.

Try it for yourself
By holding your hand in front of your face and closing one eye at a time, it is easy to see the different 2D views perceived by each eye.

35

How ears work

The human ear performs a range of functions, but how do they work?

The thing to remember when learning about the human ear is that sound is all about movement. When someone speaks or makes any kind of movement, the air around them is disturbed, creating a sound wave of alternating high and low frequency. These waves are detected by the ear and interpreted by the brain as words, tunes or sounds.

Consisting of air-filled cavities, labyrinthine fluid-filled channels and highly sensitive cells, the ear has external, middle and internal parts. The outer ear consists of a skin-covered flexible cartilage flap called the 'auricle', or 'pinna'. This feature is shaped to gather sound waves and amplify them before they enter the ear for processing and transmission to the brain. The first thing a sound wave entering the ear encounters is the sheet of tightly pulled tissue separating the outer and middle ear. This tissue is the eardrum, or tympanic membrane, and it vibrates as sound waves hit it.

Beyond the eardrum, in the air-filled cavity of the middle ear, are three tiny bones called the 'ossicles'. These are the smallest bones in the human body. Sound vibrations hitting the eardrum pass to the first ossicle, the malleus (hammer). Next the waves proceed along the incus (anvil) and then on to the (stapes) stirrup. The stirrup presses against a thin layer of tissue called the 'oval window', and this membrane enables sound waves to enter the fluid-filled inner ear.

The inner ear is home to the cochlea, which consists of watery ducts that channel the vibrations, as ripples, along the cochlea's spiralling tubes. Running through the middle of the cochlea is the organ of Corti, which is lined with minute sensory hair cells that pick up on the vibrations and generate nerve impulses that are sent to the brain as electrical signals. The brain can interpret these signals as sounds.

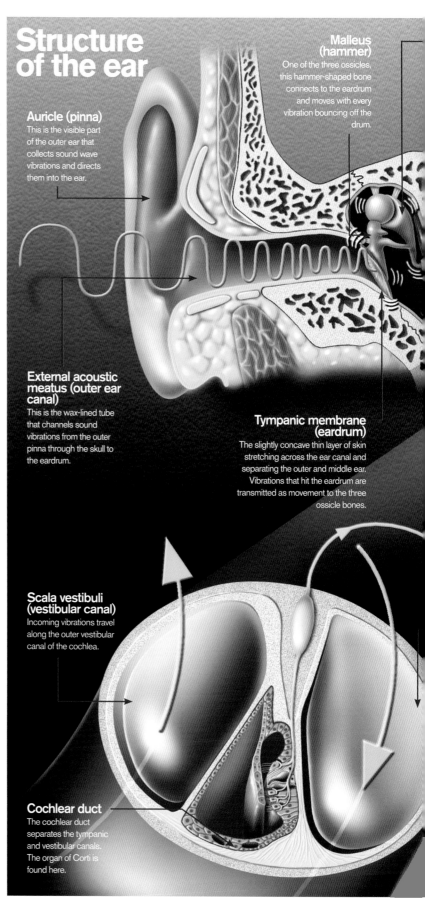

Structure of the ear

Auricle (pinna)
This is the visible part of the outer ear that collects sound wave vibrations and directs them into the ear.

Malleus (hammer)
One of the three ossicles, this hammer-shaped bone connects to the eardrum and moves with every vibration bouncing off the drum.

External acoustic meatus (outer ear canal)
This is the wax-lined tube that channels sound vibrations from the outer pinna through the skull to the eardrum.

Tympanic membrane (eardrum)
The slightly concave thin layer of skin stretching across the ear canal and separating the outer and middle ear. Vibrations that hit the eardrum are transmitted as movement to the three ossicle bones.

Scala vestibuli (vestibular canal)
Incoming vibrations travel along the outer vestibular canal of the cochlea.

Cochlear duct
The cochlear duct separates the tympanic and vestibular canals. The organ of Corti is found here.

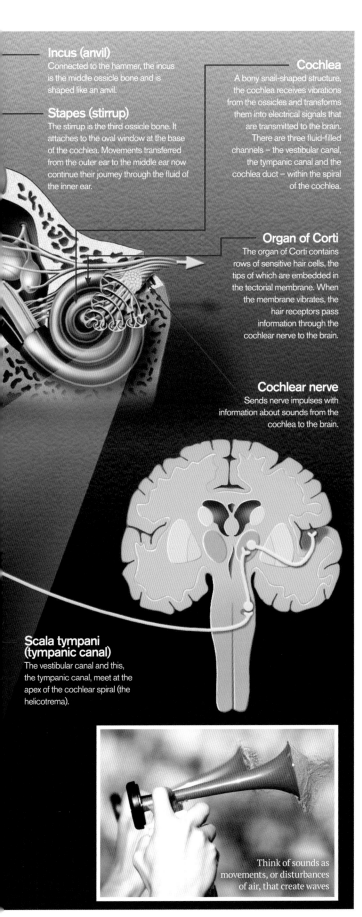

Incus (anvil)
Connected to the hammer, the incus is the middle ossicle bone and is shaped like an anvil.

Stapes (stirrup)
The stirrup is the third ossicle bone. It attaches to the oval window at the base of the cochlea. Movements transferred from the outer ear to the middle ear now continue their journey through the fluid of the inner ear.

Cochlea
A bony snail-shaped structure, the cochlea receives vibrations from the ossicles and transforms them into electrical signals that are transmitted to the brain. There are three fluid-filled channels – the vestibular canal, the tympanic canal and the cochlea duct – within the spiral of the cochlea.

Organ of Corti
The organ of Corti contains rows of sensitive hair cells, the tips of which are embedded in the tectorial membrane. When the membrane vibrates, the hair receptors pass information through the cochlear nerve to the brain.

Cochlear nerve
Sends nerve impulses with information about sounds from the cochlea to the brain.

Scala tympani (tympanic canal)
The vestibular canal and this, the tympanic canal, meet at the apex of the cochlear spiral (the helicotrema).

Think of sounds as movements, or disturbances of air, that create waves

The vestibular system

Inside the inner ear are the vestibule and semicircular canals, which feature sensory cells. From the semicircular canals and maculae, information about which way the head is moving is passed to receptors, which send electrical signals to the brain as nerve impulses.

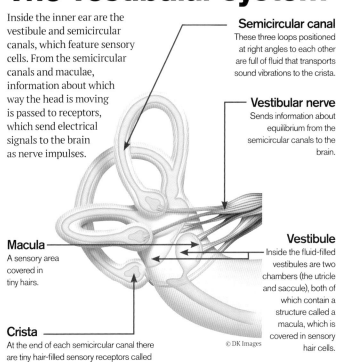

Semicircular canal
These three loops positioned at right angles to each other are full of fluid that transports sound vibrations to the crista.

Vestibular nerve
Sends information about equilibrium from the semicircular canals to the brain.

Vestibule
Inside the fluid-filled vestibules are two chambers (the utricle and saccule), both of which contain a structure called a macula, which is covered in sensory hair cells.

Macula
A sensory area covered in tiny hairs.

Crista
At the end of each semicircular canal there are tiny hair-filled sensory receptors called cristae.

© DK Images

A sense of balance

The vestibular system functions to give you a sense of which way your head is pointing in relation to gravity. It enables you to discern whether your head is upright or not, as well as helping you to maintain eye contact with stationary objects while your head is turning.

Also located within the inner ear, but less to do with sound and more concerned with the movement of your head, are the semicircular canals. Again filled with fluid, these looping ducts act like internal accelerometers that can actually detect acceleration (ie, movement of your head) in three different directions due to the positioning of the loops along different planes. Like the organ of Corti, the semicircular canals employ tiny hair cells to sense movement. The canals are connected to the auditory nerve at the back of the brain.

Your sense of balance is so complex that the area of your brain that's purely dedicated to this one role involves the same number of cells as the rest of your brain cells put together.

The surfer's semicircular canals are as crucial as his feet when it comes to staying on his board

Anatomy of the neck

Explore one of the most complex and functional areas of the human body

The human neck is a perfect blend of form and function. It has several specific tasks (eg making it possible to turn our heads to see), while serving as a conduit for other vital activities (eg connecting the mouth to the lungs).

The anatomical design of the neck would impress modern engineers. The flexibility of the cervical spine allows your head to rotate, flex and tilt many thousands of times a day.

The muscles and bones provide the strength and flexibility required, however the really impressive design comes with the trachea, oesophagus, spinal cord, myriad nerves and the vital blood vessels. These structures must all find space and function perfectly at the same time. They must also be able to maintain their shape while the neck moves.

These structures are all highly adapted to achieve their aims. The trachea is protected by a ring of strong cartilage so it doesn't collapse, while allowing enough flexibility to move when stretched. Above this, the larynx lets air move over the vocal cords so we can speak. Farther back, the oesophagus is a muscular tube which food and drink pass through en route to the stomach. Within the supporting bones of the neck sits the spinal cord, which transmits the vital nerves allowing us to move and feel. The carotid arteries and jugular veins, meanwhile, constantly carry blood to and from the brain.

How does the head connect to the neck?

They are connected at the bottom of the skull and at the top of the spinal column. The first vertebra is called the atlas and the second is called the axis. Together these form a special pivot joint that grants far more movement than other vertebrae. The axis contains a bony projection upwards, upon which the atlas rotates, allowing the head to turn. The skull sits on top of slightly flattened areas of the atlas, providing a safe platform for it to stabilise on, and allowing for nodding motions. These bony connections are reinforced with strong muscles, adding further stability. Don't forget that this amazing anatomical design still allows the vital spinal cord to pass out of the brain. The cord sits in the middle of the bony vertebrae, where it is protected from bumps and knocks. It sends out nerves at every level (starting right from the top) which actually control over most of the body.

Get it in the neck

We show the major features that are packed into this junction between the head and torso

Sympathetic trunk
These special nerves run alongside the spinal cord, and control sweating, heart rate and breathing, among other vital functions.

Cartilage
This tough tissue protects the delicate airways behind, including the larynx.

Oesophagus
This pipe connects the mouth to the stomach, and is collapsed until you swallow something, when its muscular walls stretch.

Larynx
This serves two main functions: to connect the mouth to the trachea, and to generate your voice.

Vertebra
These bones provide support to prevent the neck collapsing, hold up the skull and protect the spinal cord within.

Phrenic nerve
These important nerves come off the third, fourth and fifth neck vertebrae, and innervate the diaphragm, which keeps you breathing (without you having to think about it).

Spinal cord
Shielded by the vertebrae, the spinal cord sends motor signals down nerves and receives sensory information from all around the body.

Carotid artery
These arteries transmit oxygenated blood from the heart to the brain. There are two of them (right and left), in case one becomes blocked.

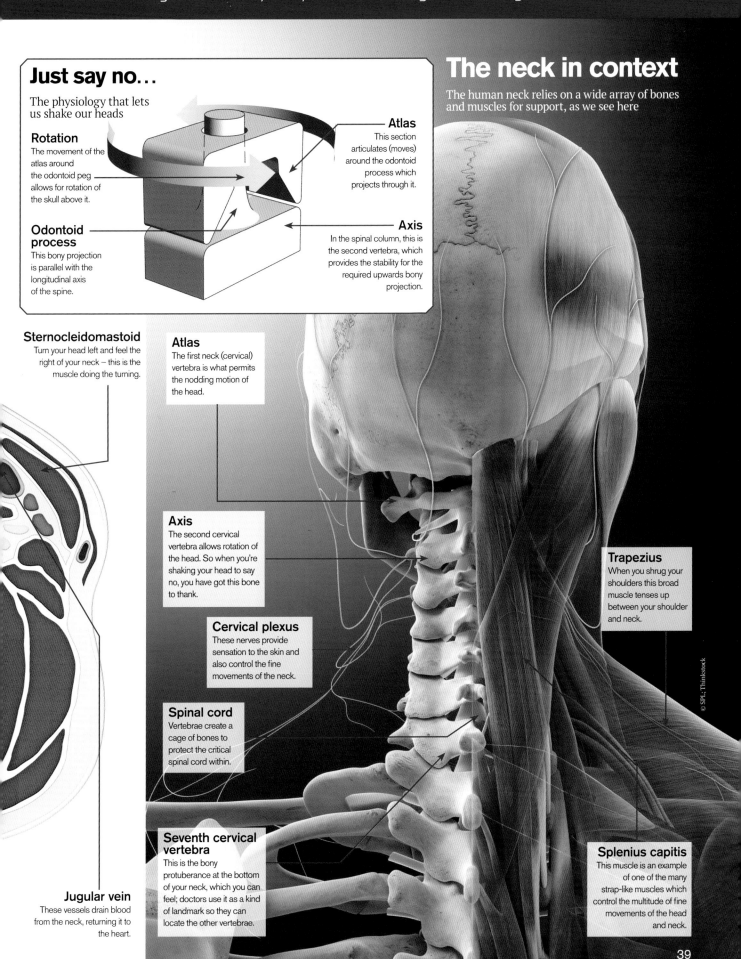

Just say no...

The physiology that lets us shake our heads

Rotation
The movement of the atlas around the odontoid peg allows for rotation of the skull above it.

Odontoid process
This bony projection is parallel with the longitudinal axis of the spine.

Atlas
This section articulates (moves) around the odontoid process which projects through it.

Axis
In the spinal column, this is the second vertebra, which provides the stability for the required upwards bony projection.

The neck in context
The human neck relies on a wide array of bones and muscles for support, as we see here

Sternocleidomastoid
Turn your head left and feel the right of your neck – this is the muscle doing the turning.

Atlas
The first neck (cervical) vertebra is what permits the nodding motion of the head.

Axis
The second cervical vertebra allows rotation of the head. So when you're shaking your head to say no, you have got this bone to thank.

Cervical plexus
These nerves provide sensation to the skin and also control the fine movements of the neck.

Spinal cord
Vertebrae create a cage of bones to protect the critical spinal cord within.

Seventh cervical vertebra
This is the bony protuberance at the bottom of your neck, which you can feel; doctors use it as a kind of landmark so they can locate the other vertebrae.

Jugular vein
These vessels drain blood from the neck, returning it to the heart.

Trapezius
When you shrug your shoulders this broad muscle tenses up between your shoulder and neck.

Splenius capitis
This muscle is an example of one of the many strap-like muscles which control the multitude of fine movements of the head and neck.

© SPL; Thinkstock

39

How the human skeleton works

Without a skeleton, we would not be able to live. It is what gives us our shape and structure and its presence allows us to operate on a daily basis. It also is a fascinating evolutionary link to all living and extinct vertebrates

The human skeleton is crucial for us to live. It keeps our shape and muscle attached to the skeleton allows us the ability to move around, while also protecting crucial organs that we need to survive. Bones also produce blood cells within bone marrow and store minerals we need released on a daily basis.

As an adult you will have around 206 bones, but you are born with over 270, which continue to grow, strengthen and fuse after birth until around 18 in females and 20 in males. Skeletons actually do vary between sexes in structure also. One of the most obvious areas is the pelvis as a female must be able to give birth, and therefore hips are comparatively shallower and wider. The cranium also becomes more robust in males due to heavy muscle attachment and a male's chin is often more prominent. Female skeletons are generally more delicate overall. However, although there are several methods, sexing can be difficult because of the level of variation we see within the species.

Bones are made up of various different elements. In utero, the skeleton takes shape as cartilage, which then starts to calcify and develop during gestation and following birth. The primary element that makes up bone, osseous tissue, is actually mineralised calcium phosphate, but other forms of tissue such as marrow, cartilage and blood vessels are also contained in the overall structure. Many individuals think that bones are solid, but actually inner bone is porous and full of little holes.

Even though cells are constantly being replaced, and therefore no cell in our body is more than 20 years old, they are not replaced with perfect, brand-new cells. The cells contain errors in their DNA and ultimately our bones therefore weaken as we age. Conditions such as arthritis and osteoporosis can often be caused by ageing and cause issues with weakening of bones and reduced movement ability.

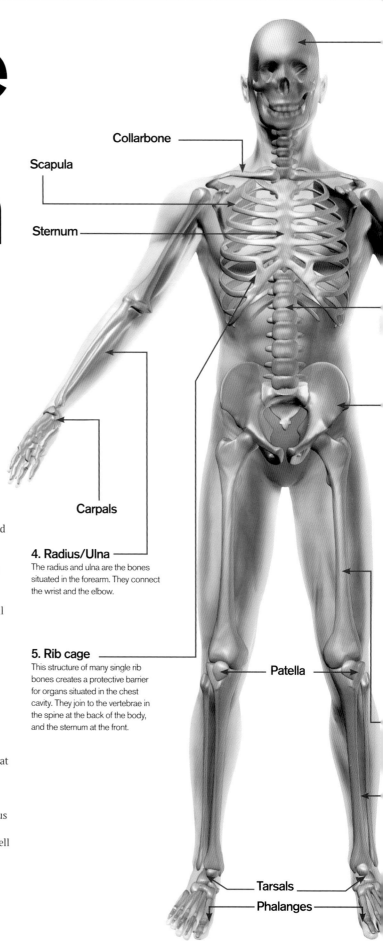

Collarbone

Scapula

Sternum

Carpals

4. Radius/Ulna
The radius and ulna are the bones situated in the forearm. They connect the wrist and the elbow.

5. Rib cage
This structure of many single rib bones creates a protective barrier for organs situated in the chest cavity. They join to the vertebrae in the spine at the back of the body, and the sternum at the front.

Patella

Tarsals

Phalanges

Inside our skeleton
How the human skeleton works and keeps us upright

1. Cranium
The cranium, also known as the skull, is where the brain and the majority of the sensory organs are located.

2. Metacarpals
The long bones in the hands are called metacarpals, and are the equivalent of metatarsals in the foot. Phalanges located close to the metacarpals make up the fingers.

3. Vertebrae
There are three main kinds of vertebrae (excluding the sacrum and coccyx) – cervical, thoracic and lumbar. These vary in strength and structure as they carry different pressure within the spine.

6. Pelvis
This is the transitional joint between the trunk of the body and the legs. It is one of the key areas in which we can see the skeletal differences between the sexes.

7. Femur
This is the largest and longest single bone in the body. It connects to the pelvis with a ball and socket joint.

8. Fibula/Tibia
These two bones form the lower leg bone and connect to the knee joint and the foot.

9. Metatarsals
These are the five long bones in the foot that aid balance and movement. Phalanges located close to the metatarsals are the bones which are present in toes.

Breaking bones

Whether it's a complete break or just a fracture, both can take time to heal properly

If you simply fracture the bone, you may just need to keep it straight and keep pressure off it until it heals. However, if you break it into more than one piece, you may need metal pins inserted into the bone to realign it or plates to cover the break in order for it to heal properly. The bone heals by producing new cells and tiny blood vessels where the fracture or break has occurred and these then rejoin up. For most breaks or fractures, a cast external to the body will be put on around the bone to take pressure off the bone to ensure that no more damage is done and the break can heal.

"The skull is actually seven separate plates when we are born, which fuse together"

Skull development

When we are born, many of our bones are still somewhat soft and are not yet fused – this process occurs later during our childhood

The primary reasons for the cranium in particular not to be fully fused at birth is to allow the skull to flex as the baby is born and also to allow the extreme rate of growth that occurs in the first few years of childhood following birth. The skull is actually in seven separate plates when we are born and over the first two years these pieces fuse together slowly and ossify. The plates start suturing together early on, but the anterior fontanel – commonly known as the soft spot – will take around 18 months to fully heal. Some other bones, such as the five bones located in the sacrum, don't fully fuse until late teens or early twenties, but the cranium becomes fully fused by around age two.

3 skulls © DK Images

How our joints work
The types of joints in our body explained

1. Ball and socket joints
Both the hip and the shoulder joints are ball and socket joints. The femur and humerus have ball shaped endings, which turn in a cavity to allow movement.

2. Vertebrae
Vertebrae fit together to support the body and allow bending movements. They are joined by cartilage and are classified as semi-mobile joints.

3. Skull sutures
Although not generally thought of as a 'joint', all the cranial sutures present from where bones have fused in childhood are in fact immoveable joints.

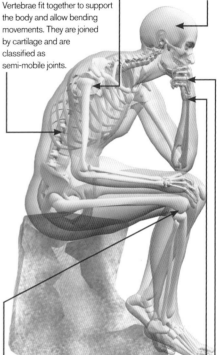

4. Hinged joints
Both elbows and knees are hinged joints. These joints only allow limited movement in one direction. The bones fit together and are moved by muscles.

5. Gliding joints
Some movement can be allowed when flat bones 'glide' across each other. The wrist bones – the carpals – operate like this, moved by ligaments.

6. Saddle joints
The only place we see this joint in humans is the thumb. Movement is limited in rotation, but the thumb can move back, forward and to the sides.

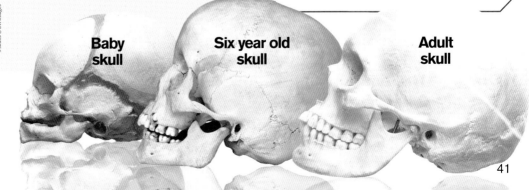

Baby skull   **Six year old skull**   **Adult skull**

41

Breaking it down

How the various bones fit together to form the skull

Frontal bone
The single bone that forms the forehead, it is often considered a facial bone despite being a calvarial bone.

Maxilla
Comprising part of the upper jaw and hard palate, the maxilla also forms part of the nose and eye socket.

Zygoma
This bony arch spans from the cheek to just above the ear canal.

Parietal bones
The parietal bones form most of the upper lateral side of the skull, joining together at the top.

Occipital bone
Located at the back of the skull, this section of bone contains openings for the spinal cord, nerves and vessels.

Temporal bone
Divided into four parts, the temporal bone supports the temple and houses the structures that enable us to hear.

Sphenoid
The complex sphenoid is a crucial bone, as it joins with almost every other skull bone.

Mandible
The only moveable bone of the skull, the lower jaw is the largest and strongest one too.

The human skull

Understand the complex structure that supports our brain and facial tissues

The human skull is made up of 22 bones that fall into two primary groups: the cranium, which consists of eight 'cranial' bones, and the facial skeleton, which consists of 14 facial bones. These bones are joined together by fibrous joints known as sutures. Unique to the skull, once these joints have fused together by the age of around 30 to 40, they are immovable.

The cranium consists of a roof part – known as the calvarium – and a complex base part. The calvarium helps to cover the cranial cavity, which the brain occupies, along with flat bones at the top and sides. The base of the brain case is divided into large spaces, and has various openings for the passages of cranial nerves, blood vessels and the spinal cord.

The facial skeleton provides support for the soft facial tissues, and its bones fuse together to house the orbits of the eyes, nasal and oral cavities, in addition to the sinuses. Only one of the skull's 22 bones is moveable, and that is the lower jaw.

As you can see in the diagram above, the skull is a complex structure, but then its main roles are to protect the brain and support the face, so this comes as no surprise.

The skull does an impressive job of protecting superficial muscles, nerves and blood vessels

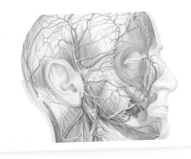

Why do babies have a 'soft spot'?

The open spaces between the bones of a baby's skull where the sutures intersect are called fontanelles. Covered in a protective membrane, there are two kinds of fontanelles: the anterior fontanelle – also known as the 'soft spot' – and the posterior fontanelle. The anterior fontanelle is where the two frontal and two parietal bones meet, and this area stays soft until the age of around two. The posterior fontanelle is where the two parietal bones and the occipital bone meet, and this area usually closes when a baby is a few months old.

The formation occurs as a means of helping the baby's head fit through the birth canal. The reason they remain open for some time is to enable the brain to develop and grow. It's important that the fontanelles don't close up too early – a process known as craniosynostosis – as this can result in various medical problems.

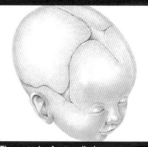

The anterior fontanelle is positioned at the front of the skull

How do humans speak?

The vocal cords and larynx in particular have evolved over time to enable humans to produce a dramatic range of sounds in order to communicate – but how do they work?

V ocal cords, also known as vocal folds, are situated in the larynx, which is placed at the top of the trachea. They are layers of mucous membranes that stretch across the larynx and control how air is expelled from the lungs in order to make certain sounds. The primary usage of vocal cords within humans is in order to be able to communicate with eachother and it is hypothesised that human vocal cords actually developed to the extent we see now to facilitate advanced levels of communication in response to the formation of social groupings during phases of primate, and specifically human, evolution.

As air is expelled from the lungs, the vocal folds vibrate and collide to produce a range of sounds. The type of sound emitted is affected by exactly how the folds collide, move and stretch as air passes over them. An individual 'fundamental frequency' is determined by the length, size and tension of their vocal cords. Movement of the vocal folds is controlled by the vagus nerve, and sound is then further fine-tuned to form words and sounds that we can recognise by the larynx, tongue and lips. Fundamental frequency in males averages at 125Hz, and at 210Hz in females. Children have a higher average pitch at around 300Hz.

Differences between male and female vocal cords

Male voices are often much lower than female voices. This is primarily due to the different size of vocal folds present in each sex, with males having larger folds that create a lower pitched sound, and females having smaller folds that create a higher pitch sound. The average size for male vocal cords are between 17 and 25mm, and females are normally between 12.5 and 17.5mm. From the range in size, however, males can be seen to have quite high pitch voices, and females can have quite low pitch voices.

The other major biological difference that effects pitch is that males generally have a larger vocal tract, which can further lower the tone of their voice independent of vocal cord size. The pitch and tone of male voices has been studied in relation to sexual success, and individuals with lower voices have been seen to be more successful in reproduction. The reason proposed for this is that a lower tone voice may indicate a higher level of testosterone present in a male.

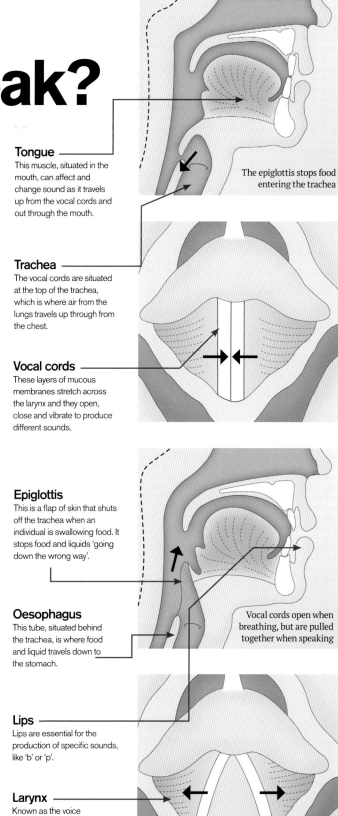

Tongue
This muscle, situated in the mouth, can affect and change sound as it travels up from the vocal cords and out through the mouth.

The epiglottis stops food entering the trachea

Trachea
The vocal cords are situated at the top of the trachea, which is where air from the lungs travels up through from the chest.

Vocal cords
These layers of mucous membranes stretch across the larynx and they open, close and vibrate to produce different sounds.

Epiglottis
This is a flap of skin that shuts off the trachea when an individual is swallowing food. It stops food and liquids 'going down the wrong way'.

Oesophagus
This tube, situated behind the trachea, is where food and liquid travels down to the stomach.

Vocal cords open when breathing, but are pulled together when speaking

Lips
Lips are essential for the production of specific sounds, like 'b' or 'p'.

Larynx
Known as the voice box, this protects the trachea and is heavily involved in controlling pitch and volume. The vocal cords are situated within the larynx.

The human spine

The human spine is made up of 33 vertebrae, but how do they support our bodies while also allowing us such flexibility?

The human spine is made up of 33 vertebrae, 24 of which are articulated (flexible) and nine of which normally become fused in maturity. They are situated between the base of the skull to the pelvis, where the spine trails off into the coccyx – an evolutionary remnant of a tail our ancestors would have displayed.

The primary functions of the vertebrae that make up the spine are to support the torso and head, which protect vital nerves and the spinal cord and allow the individual to move. By sitting closely together, separated only by thin intervertebral discs which work as ligaments and effectively form joints between the bones, the vertebrae form a strong pillar structure which holds the head up and allows for the body to remain upright. It also produces a base for ribs to attach to and to protect vital internal organs in the human body.

Vertebrae are not all fused together because of the need to move, and the vertebrae themselves are grouped into five types – cervical, thoracic, lumbar, sacral and coccygeal. The sacral vertebrae fuse during maturity (childhood and teenage years) and become solid bones towards the base of the spine. The coccygeal vertebrae will fuse in some cases, but studies have shown that often they actually remain separate. Collectively they are referred to as the coccyx (tail bone). The rest of the vertebrae remain individual and allow discs between them to move them to move in various directions without wearing the bones down. The cervical vertebrae in the neck allow particularly extensive movement, allowing the head to move up and down and side to side. The thoracic are far more static, with ties to the rib cage resisting much movement. The lumbar vertebrae allow modest side-to-side movement and rotation. A particular feature of the spine is how it is actually curved to allow distribution of the body's weight, to ensure no one vertebrae takes the full impact.

Cervical vertebrae

These are the smallest of the articulating vertebrae, and support the head and neck. There are seven vertebrae, with C1, C2 and C7's structures quite unique from the others. They sit between the skull and thoracic vertebrae.

C1 (atlas)

This is the vertebrae which connects the spinal column with the skull. It is named 'atlas' after the legend of Atlas who held the entire world on his shoulders.

C2 (axis)

C2 is the pivot for C1 (atlas), and nearly all movement for shaking your head will occur at this joint – the atlanto-axial joint.

Thoracic vertebrae

The thoracic vertebrae are the intermediately sized vertebrae. They increase in size as you move down the spine, and they supply facets for ribs to attach to – this is how they are primarily distinguished.

Intervertebral discs

These discs form a joint between each vertebrae and, effectively, work as ligaments while also serving as fantastic shock absorbers. They facilitate movement and stop the bones rubbing together.

Spine curvature

As you look at the human spine, you can see some distinct curves. The primary reasons for these are to help distribute weight throughout the spine and support certain aspects of the body. The curve most familiar to us is the lumbar curve, between the ribs and pelvis. This develops when we start to walk at about 12-18 months and helps us with weight distribution during locomotion. Prior to this we develop the cervical curve, which allows us to support the weight of our head at around three-four months, and two smaller less-obvious curves in the spine (the thoracic and pelvic curves) are developed during gestation.

Spinal cords and nerves

The human spinal cord is an immensely complex structure made up of nerve cells and a large amount of supporting, protective tissue. It splits into 31 different sections and stretches 43-45cm, down from the brain to between the first and second lumbar vertebrae. Although more commonly referred to in respect of the brain, there is both white and grey matter present in the centre of the spinal cord. White matter contains axons tracts surrounded by fats, and blood vessels to protect them. The grey matter contains more of the neural cell bodies, such as dendrites, more axons and glial cells.

Spinal cord injuries are normally caused by trauma. If the trauma causes intervertebral discs and vertebrae to break, they can pierce the spinal cord, which can result in loss of feeling. Cord severance may result in paralysis.

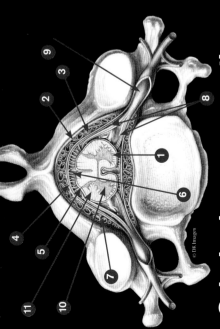

Articulated vertebrae enable maximum flexibility

How is the skull attached to the spine?

The skull is connected to the spine by the atlanto-occipital joint, which is created by C1 (atlas) and the occipital bone situated at the base of the cranium (skull). This unique vertebra has no 'body' and actually looks more like a ring than any other vertebra. It sits at the top of the cervical vertebrae and connects with the occipital bone via an ellipsoidal joint, allowing movement such as nodding or rotation of the head. An ellipsoidal joint is where an ovoid connection (in this case the occipital bone) is placed into an elliptical cavity (C1 vertebra). The rest of the cervical vertebrae also work to support the weight of the head.

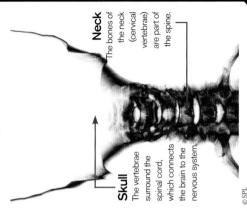

Neck
The bones of the neck (cervical vertebrae) are part of the spine.

Skull
The vertebrae surround the spinal cord, which connects the brain to the nervous system.

© SPL

Lumbar vertebrae
Lumbar vertebrae are the largest of the vertebrae and the strongest, primarily because they withstand the largest pressures. Compared with other vertebrae they are more compact, lacking facets on the sides of the vertebrae.

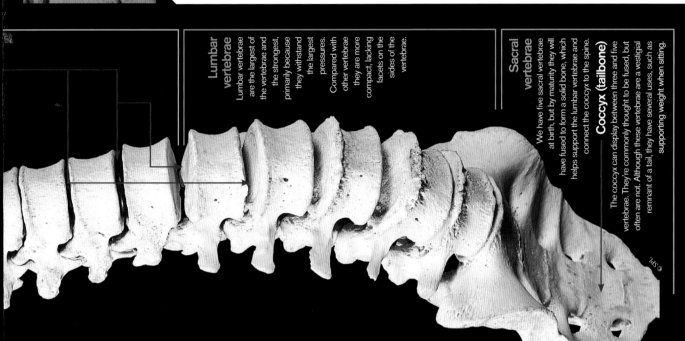

© SPL

Sacral vertebrae
We have five sacral vertebrae at birth, but by maturity they will have fused to form a solid bone, which helps support the lumbar vertebrae and connect the coccyx to the spine.

Coccyx (tailbone)
The coccyx can display between three and five vertebrae. They're commonly thought to be fused, but often are not. Although these vertebrae are a vestigial remnant of a tail, they have several uses, such as supporting weight when sitting.

© DK Images

Spinal column cross-section

1. Spinal cord
This is an immensely important pathway for information to transfer between the brain and the body's nervous system. It is heavily protected by tissue and vertebrae, as any damage to it can be fatal.

2. Epidural space
This is the space between the outer protective tissue layer, dura mater and the bone. It is filled with adipose tissue (fat), while also playing host to numerous blood vessels.

3. Dura mater
This is the tough outer layer of tissue that protects the spinal cord. The three layers of protection between the vertebrae and the spinal cord are called the spinal meninges.

4. Arachnoid mater
Named for its spider web appearance, this is the second layer of the tissue protection provided for the spinal cord.

5. Pia mater
This thin, delicate layer sits immediately next to the spinal cord.

6. Subarachnoid space
This is the space between the pia mater and the arachnoid mater, which is filled with cerebrospinal fluid.

7. Blood vessels
Four arteries, which form a network called the Circle of Willis, deliver oxygen-rich blood to the brain. The brain's capillaries form a lining called the 'blood-brain barrier', which controls blood flow to the brain.

8. Dorsal and ventral roots
These connect the spinal nerves to the spinal cord, allowing transition of information between the brain and the body.

9. Spinal nerves
Humans have 31 pairs of spinal nerves all aligned with individual vertebrae, and these communicate information from around the body to the spinal cord. They carry all types of information – motor, sensory and so on – and are commonly referred to as 'mixed spinal nerves'.

10. Grey matter
Within the horn-like shapes in the centre of the spinal cord, sit most of the important neural cell bodies. They are protected in many ways, including by the white matter.

11. White matter
This area that surrounds the grey matter holds axon trails, but is primarily made up of lipid tissue (fats) and blood vessels.

Joints

For bones to function together, they are linked by joints

Some bones, like those in the skull, do not need to move, and are permanently fused together with mineral sutures. These fixed joints provide maximum stability. However, most bones need flexible linkages. In some parts of the skeleton, partial flexibility is sufficient, so all that the bones require is a little cushioning to prevent rubbing. The bones are joined by a rigid, gel-like tissue known as cartilage, which allows for a small range of compression and stretching. These types of joints are present where the ribs meet the sternum, providing flexibility when breathing, and between the stacked vertebrae of the spinal column, allowing it to bend and flex without crushing the spinal cord.

Most joints require a larger range of movement. Covering the ends of the bones in cartilage provides shock absorption, but for them to move freely in a socket, the cartilage must be lubricated to make it slippery and wear-proof. At synovial joints, the ends of the two bones are encased in a capsule, covered on the inside by a synovial membrane, which fills the joint with synovial fluid, allowing the bones to slide smoothly past one another.

There are different types of synovial joint, each with a different range of motion. Ball-and-socket joints are used at the shoulder and hip, and provide a wide range of motion, allowing the curved surface at the top end of each limb to slide inside a cartilage covered cup. The knees and elbows have hinge joints, which interlock in one plane, allowing the joint to open and close. For areas that need to be flexible, but do not need to move freely, such as the feet and the palm of the hand, gliding joints allow the bones to slide small distances without rubbing.

Bone joints

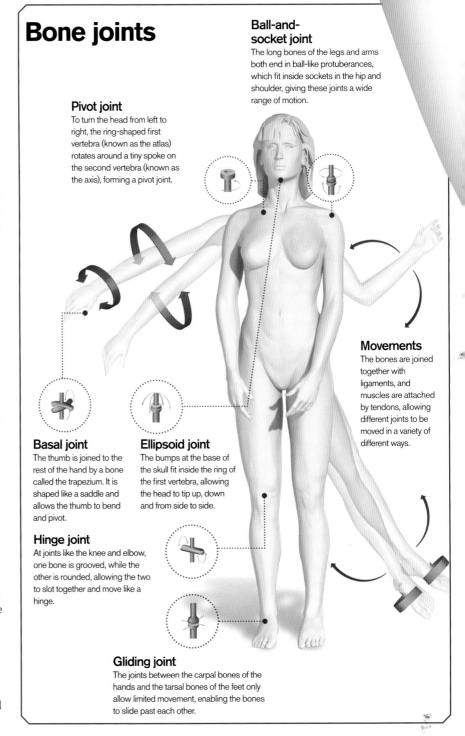

Ball-and-socket joint
The long bones of the legs and arms both end in ball-like protuberances, which fit inside sockets in the hip and shoulder, giving these joints a wide range of motion.

Pivot joint
To turn the head from left to right, the ring-shaped first vertebra (known as the atlas) rotates around a tiny spoke on the second vertebra (known as the axis), forming a pivot joint.

Movements
The bones are joined together with ligaments, and muscles are attached by tendons, allowing different joints to be moved in a variety of different ways.

Basal joint
The thumb is joined to the rest of the hand by a bone called the trapezium. It is shaped like a saddle and allows the thumb to bend and pivot.

Ellipsoid joint
The bumps at the base of the skull fit inside the ring of the first vertebra, allowing the head to tip up, down and from side to side.

Hinge joint
At joints like the knee and elbow, one bone is grooved, while the other is rounded, allowing the two to slot together and move like a hinge.

Gliding joint
The joints between the carpal bones of the hands and the tarsal bones of the feet only allow limited movement, enabling the bones to slide past each other.

Hypermobility
Some people tend to have particularly flexible joints and a much larger range of motion. This is sometimes known as being 'double jointed.' It is thought to result from the structure of the collagen in the joints, the shape of the end of the bones, and the tone of the muscles around the joint.

Mobile
The synovial joints are the most mobile in the body. The ends of the bones are linked by a capsule that contains a fluid lubricant, allowing the bones to slide past one another. Synovial joints come in different types, including ball-and-socket, hinge, and gliding.

Semi-mobile
Cartilaginous joints do not allow free motion, but cushion smaller movements. Instead of a lubricated capsule, the bones are joined by fibrous or hyaline cartilage. The linkage acts as a shock absorber, so the bones can move apart and together over small distances.

Fixed
Some bones do not need to move relative to one another and are permanently fused. For example the cranium starts out as separate pieces, allowing the foetal head to change shape to fit through the birth canal, but fuses after birth to encase the brain in a solid protective skull.

Why our joints crack

The synovial fluid used to lubricate the joints contains dissolved gasses. The fluid is sealed within a capsule, so if the joint is stretched, the capsule also stretches, creating a vacuum as the pressure changes, and pulling the gas out of solution and into a bubble, which pops, producing a cracking sound.

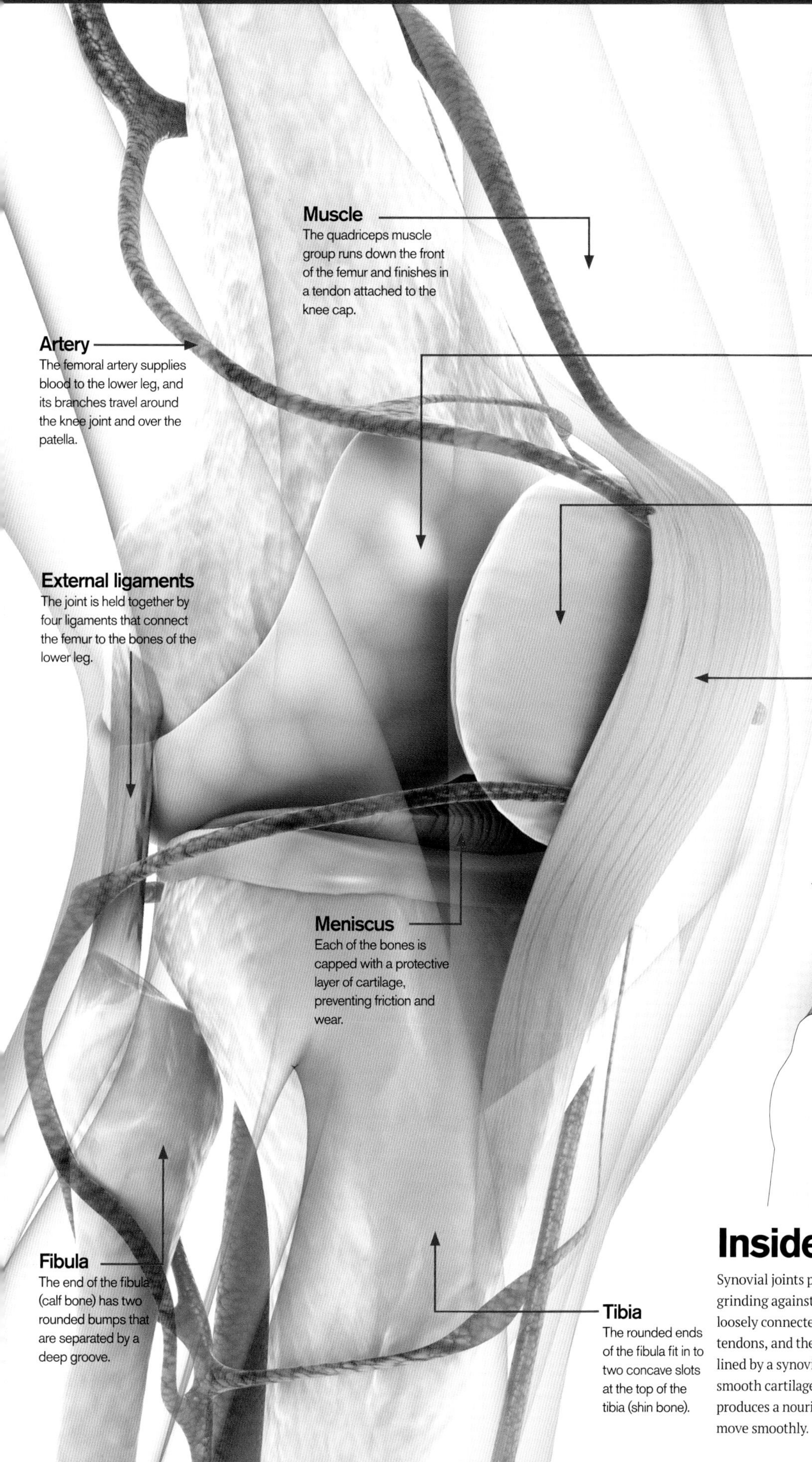

Muscle
The quadriceps muscle group runs down the front of the femur and finishes in a tendon attached to the knee cap.

Artery
The femoral artery supplies blood to the lower leg, and its branches travel around the knee joint and over the patella.

External ligaments
The joint is held together by four ligaments that connect the femur to the bones of the lower leg.

Synovial membrane
The membrane surrounding the interior of the joint produces a lubricant called synovial fluid.

Knee cap
The patella prevents the tendons at the front of the leg from wearing away at the joint.

Patellar ligament
The patellar ligament connects the kneecap to both the quadriceps in the thigh and the tibia in the lower leg.

Meniscus
Each of the bones is capped with a protective layer of cartilage, preventing friction and wear.

Fibula
The end of the fibula (calf bone) has two rounded bumps that are separated by a deep groove.

Tibia
The rounded ends of the fibula fit in to two concave slots at the top of the tibia (shin bone).

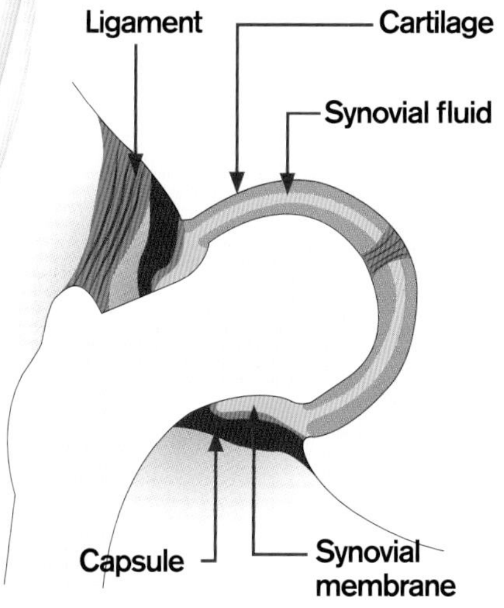

Ligament　**Cartilage**

Synovial fluid

Capsule　**Synovial membrane**

Inside a joint

Synovial joints prevent mobile areas of the skeleton from grinding against one another as they move. The two bones are loosely connected by strips of connective tissue called tendons, and the two ends are encased in a capsule that is lined by a synovial membrane. The bones are covered in smooth cartilage to prevent abrasion and the membrane produces a nourishing lubricant to ensure the joint is able to move smoothly.

How do muscles work?

Muscles are essential for us to operate on a daily basis, but how are they structured and how do they keep us moving

A muscle is a group of tissue fibres that contract and release to control movements within the body. We have three different types of muscles in our bodies – smooth muscle, cardiac muscle and skeletal muscle.

Skeletal muscle, also known as striated muscle, is what we would commonly perceive as muscle, this being external muscles that are attached to the skeleton, such as biceps and deltoids. These muscles are connected to the skeleton with tendons. Cardiac muscle concerns the heart, which is crucial as it pumps blood around the body, supplying oxygen and ultimately energy to muscles, which allows them to operate. Smooth muscle, which is normally sheet muscle, is primarily involved in muscle contractions such as bladder control and oesophagus movements. These are often referred to as involuntary as we have little or no control over these muscles' actions.

Muscles control most functions within our bodies; release of waste products, breathing, seeing, eating and movement to name but a few. Actual muscle structure is quite complex, and each muscle is made up of numerous fibres which work together to give the muscle strength. Muscles increase in effectiveness and strength through exercise and growth and the main way this occurs is through small damage caused by each repetition of a muscle movement, which the body then automatically repairs and improves.

More than 640 muscles are actually present across your entire body working to enable your limbs to work, control bodily functions and shape the body as a whole.

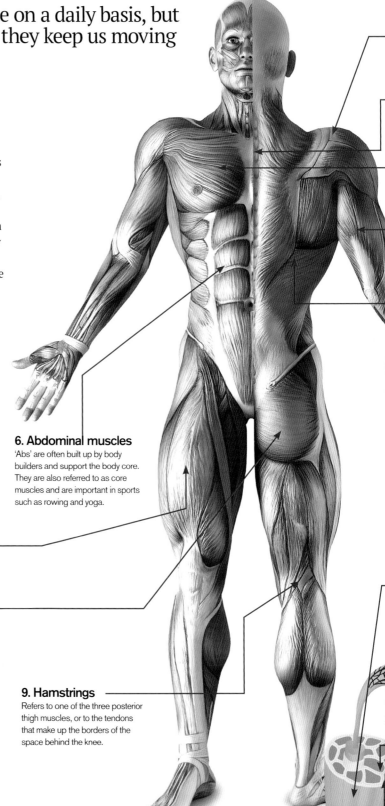

6. Abdominal muscles
'Abs' are often built up by body builders and support the body core. They are also referred to as core muscles and are important in sports such as rowing and yoga.

7. Quadriceps
The large fleshy muscle group covering the front and sides of the thigh.

8. Gluteus maximus
The biggest muscle in the body, this is primarily used to move the thighs back and forth.

9. Hamstrings
Refers to one of the three posterior thigh muscles, or to the tendons that make up the borders of the space behind the knee.

"More than 300 individual muscles are present across your body to enable your limbs to work"

1. **Deltoids**
These muscles stretch across the shoulders and aid lifting.

2. **Trapezius**
Large, superficial muscle at the back of the neck and the upper part of the thorax, or chest.

3. **Pectoralis major**
Commonly known as the 'pecs', this group of muscles stretch across the chest.

4. **Biceps/triceps**
These arm muscles work together to lift the arm up and down. Each one contracts, causing movement in the opposite direction to the other.

5. **Latissmus dorsi**
Also referred to as the 'lats', these muscles are again built up during weight training and are used to pull down objects from above.

What affects our muscle strength?

How strong we are is a combination of nature and nurture

Muscle strength refers to the amount of force that a muscle can produce, while operating at maximum capacity, in one contraction. Size and structure of the muscle is important for muscle strength, with strength being measured in several ways. Consequently, it is hard to definitively state which muscle is actually strongest.

We have two types of muscle fibre – one that supports long, constant usage exerting low levels of pressure, and one that supports brief, high levels of force. The latter is used during anaerobic activity and these fibres respond better to muscle building.

Genetics can affect muscle strength, as can usage, diet and exercise regimes. Contractions of muscles cause injuries in the muscle fibres and it is the healing of these that actually create muscle strength as the injuries are repaired and overall strengthen the muscle.

"Tendons attach muscles such as biceps to bones, allowing muscles to move elements of our body"

How does the arm flex?

Biceps and triceps are a pair of muscles that work together to move the arm up and down. As the bicep contracts, the triceps will relax and stretch out and consequently the arm will move upwards. When the arm needs to move down, the opposite will occur – with the triceps contracting and the bicep relaxing and being forcibly stretched out by the triceps. The bicep is so named a flexor as it bends a joint, and triceps would be the extensor as it straightens the joint out. Neither of these muscles can push themselves straight, they depend on the other to oppose their movements and stretch them out. Many muscles therefore work in pairs, so-called antagonistic muscles.

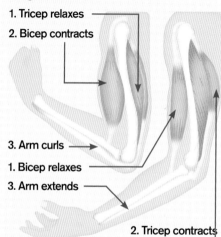

1. Tricep relaxes
2. Bicep contracts
3. Arm curls
1. Bicep relaxes
3. Arm extends
2. Tricep contracts

What is a pulled muscle, and how does it happen?

They hurt like crazy so here's why it's important to warm up

A pulled muscle is a tear in muscle fibres. Sudden movements commonly cause pulled muscles, and when an individual has not warmed up appropriately before exercise or is unfit, a tear can occur as the muscle is not prepared for usage. The most common muscle to be pulled is the hamstring, which stretches from the buttock to the knee. A pulled muscle may result in swelling and the pain can last for several days before the fibres can repair themselves. To prevent pulling muscles, warming up is advised before doing any kind of physical exertion.

What are muscles made up of?

Muscles are made up of numerous cylindrical fibres, which work together to contract and control parts of the body. Muscle fibres are bound together by the perimysium into small bundles, which are then grouped together by the epimysium to form the actual muscle.

Blood vessels and nerves also run through the connective tissue to give energy to the muscle and allow feedback to be sent to the brain. Tendons attach muscles such as biceps and triceps to bones, allowing muscles to move elements of our body as we wish.

Epimysium
The external layer that covers the muscle overall and keeps the bundles of muscle fibres together.

Blood vessel
This provides oxygen and allows the muscle to access energy for muscle operation.

Perimysium
This layer groups together muscle fibres within the muscle.

Filaments
Myofibrils are constructed of filaments, which are made up of the proteins actin and myosin.

Endomysium
This layer surrounds each singular muscle fibre and keeps the myofibril filaments grouped together.

Tendon
These attach muscle to bones, which in turn enables the muscles to move parts of the body around (off image).

Myofibril
Located within the single muscle fibres, myofibrils are bundles of actomyosin filaments. They are crucial for contraction.

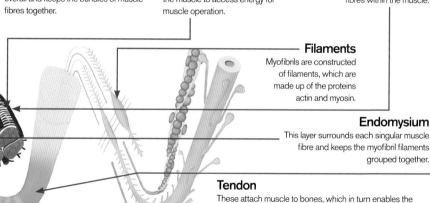

Skin senses

How does your skin pick up signals from the outside world?

The skin uses different detectors (called sensory receptors) to tell the difference between pressure and pain, hot and cold, and a light brush versus a hard poke. These receptors are specially adapted nerve endings, and many are wrapped in layers of tissue, helping them to function in various ways. They are located in different numbers across your body, with more in your hands, feet and lips than anywhere else. Some respond quickly before stopping, allowing you to grow accustomed to sensations that don't need constant monitoring, like the feeling of your clothes against your skin. Others are slow to stop signalling, so you remain aware of the sensation.

Under the skin

Different receptors take responsibility for detecting different aspects of touch

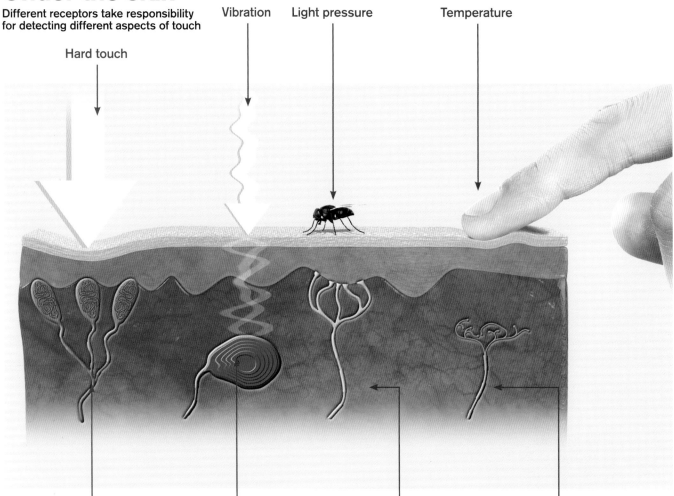

Hard touch

Vibration

Light pressure

Temperature

Meissner corpuscles
These sit in the bumps and ridges beneath the top layer of skin on the hands and the soles of the feet. They are sensitive to pressure and vibrations, which is useful for detecting texture.

Pacinian corpuscles
These are the largest of the sensory detectors in the skin. They sit deep below the surface, and are equipped to respond to fast vibration, deep pressure and tickles.

Merkel's discs
There are lots of these in the fingertips and lips, and they are good at sensing light pressure. They are particularly important for working out the shape and texture of objects.

Free nerve endings
Lots of the nerve endings under the skin aren't buried inside specific structures. These 'free' endings mainly detect pain, but some pick up on sensations such as pressure or temperature.

Under the skin

Find out more about the largest organ in your body…

Our skin is the largest organ in our bodies with an average individual skin's surface area measuring around two square metres and accounting for up to 16 per cent of total body weight. It is made up of three distinct layers. These are the epidermis, the dermis and the hypodermis and they all have differing functions. Humans are rare in that we can see these layers distinctly.

The epidermis is the top, waterproofing layer. Alongside helping to regulate temperature of the body, the epidermis also protects against infection as it stops pathogens entering the body. Although generally referred to as one layer, it is actually made up of five. The top layers are actually dead keratin-filled cells which prevent water loss and provide protection against the environment, but the lower levels, where new skin cells are produced, are nourished by the dermis. In other species, such as amphibians, the epidermis consists of only live skin cells. In these cases, the skin is generally permeable and actually may be a major respiratory organ.

The dermis has the connective tissue and nerve endings, contains hair follicles, sweat glands, lymphatic and blood vessels. The top layer of the dermis is ridged and interconnects securely with the epidermis.

Although the hypodermis is not actually considered part of the skin, its purpose is to connect the upper layers of skin to the body's underlying bone and muscle. Blood vessels and nerves pass through this layer to the dermis. This layer is actually crucial for all of the skins temperature regulation, as it contains 50 per cent of a healthy adult's body fat in subcutaneous tissue. These kinds of layers are not often seen in other species, humans being one of few that you can see the distinct layers within the skin. Not only does the skin offer protection for muscle, bone and internal organs, but it is our protective barrier against the environment. Temperature regulation, insulation, excretion of sweat and sensation are just a few more functions of skin.

1. Epidermis
This is the top, protective layer. It is waterproof and protects the body against UV light, disease and dehydration among other things.

3. Nerve ending
Situated within the dermis, nerve endings allow us to sense temperature, pain and pressure. This gives us information on our environment and stops us hurting ourselves.

5. Subcutaneous tissue
The layer of fat found in the hypodermis that is present to prevent heat loss and protect bone and muscle from damage. It is also a reserve energy source.

2. Dermis
The layer that nourishes and helps maintain the epidermis, the dermis houses hair roots, nerve endings and sweat glands.

4. Pore
Used for temperature regulation, this is where sweat is secreted to cool the body down when it is becoming too hot.

© DK Images

How your skin works
The skin is made of many more elements than most people imagine

The human heartbeat

How one of your hardest-working muscles keeps your blood pumping

Your heart began to beat when you were a four-week-old foetus in the womb. Over the course of the average lifetime, it will beat over 2 billion times.

The heart is composed of four chambers separated into two sides. The right side receives deoxygenated blood from the body, and pumps it towards the lungs, where it picks up oxygen from the air you breathe. The oxygenated blood returns to the left side of the heart, where it is sent through the circulatory system, delivering oxygen and nutrients around the body.

The pumping action of the heart is coordinated by muscular contractions that are generated by electrical currents. These currents regularly trigger cardiac contractions known as systole. The upper chambers, or atria, which receive blood arriving at the heart, contract first. This forces blood to the lower, more muscular chambers, known as ventricles, which then contract to push blood out to the body. Following a brief stage where the heart tissue relaxes, known as diastole, the cycle begins again.

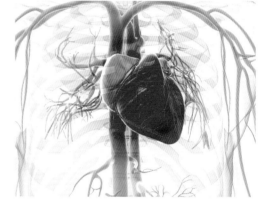

The heart consists of four chambers, separated into two sides

The cardiac cycle
A single heartbeat is a series of organised steps that maximise blood-pumping efficiency

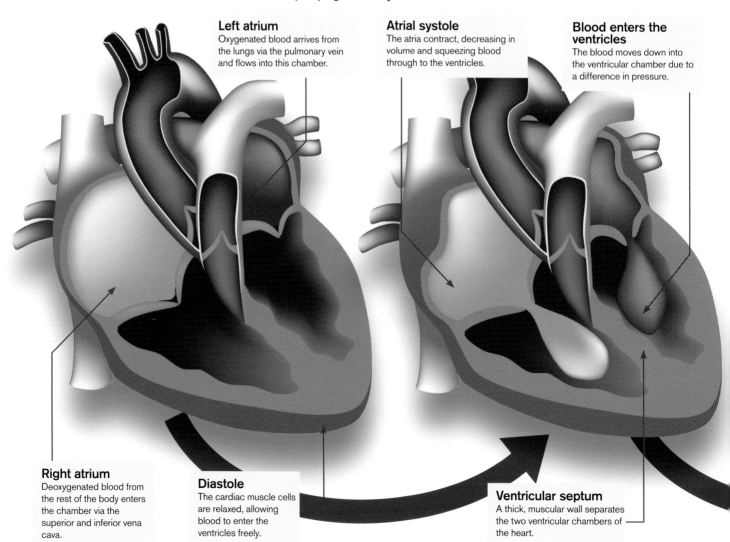

Left atrium
Oxygenated blood arrives from the lungs via the pulmonary vein and flows into this chamber.

Atrial systole
The atria contract, decreasing in volume and squeezing blood through to the ventricles.

Blood enters the ventricles
The blood moves down into the ventricular chamber due to a difference in pressure.

Right atrium
Deoxygenated blood from the rest of the body enters the chamber via the superior and inferior vena cava.

Diastole
The cardiac muscle cells are relaxed, allowing blood to enter the ventricles freely.

Ventricular septum
A thick, muscular wall separates the two ventricular chambers of the heart.

"Over the course of the average lifetime, the heart will beat over 2 billion times"

Fight or flight

A heartbeat begins at the sinoatrial node, a bundle of specialised cells in the right atrium. This acts as a natural pacemaker by generating an electrical current that moves throughout the heart, causing it to contract. When you are at rest, this happens between 60 to 100 times per minute on average. Under stressful situations however, such as an encounter with a predator, your brain will automatically trigger a 'fight or flight' response.

This results in the release of adrenaline and noradrenaline hormones that change the conductance of the sinoatrial node, increasing heart rate, and so providing the body with more available nutrients to either fight for survival or run for the hills.

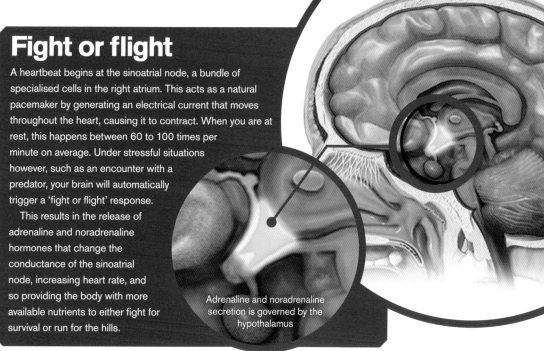

Adrenaline and noradrenaline secretion is governed by the hypothalamus

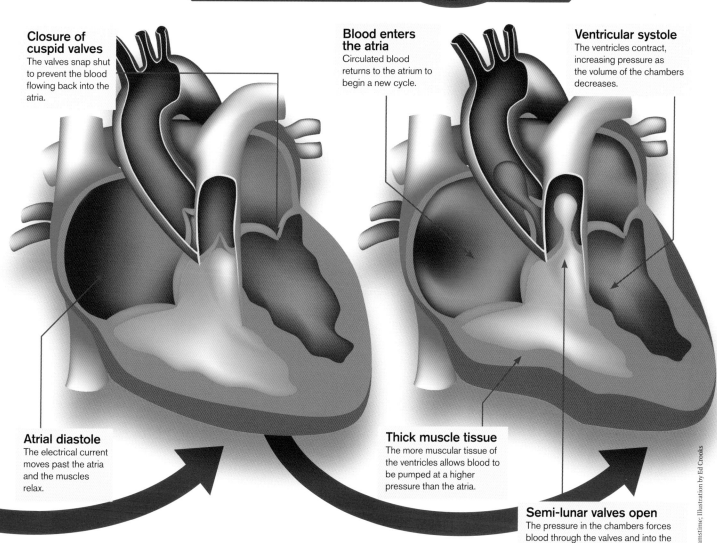

Closure of cuspid valves
The valves snap shut to prevent the blood flowing back into the atria.

Blood enters the atria
Circulated blood returns to the atrium to begin a new cycle.

Ventricular systole
The ventricles contract, increasing pressure as the volume of the chambers decreases.

Atrial diastole
The electrical current moves past the atria and the muscles relax.

Thick muscle tissue
The more muscular tissue of the ventricles allows blood to be pumped at a higher pressure than the atria.

Semi-lunar valves open
The pressure in the chambers forces blood through the valves and into the aorta and pulmonary artery.

© Dreamstime; Illustration by Ed Crooks

Anatomy of facial expressions

Does it really take more muscles to frown than to smile?

The 43 muscles of the face sit just under the skin. At one end, they are attached to bone, or sheets of tissue known as fascia, and, unlike any other muscles in the body, they join directly to the skin at the other end.

We can sort our facial muscles into three groups: the orbital group, the nasal group and the oral group. Together, they enable us to make four core expressions: happy, sad, afraid and angry, and over 20 combined expressions.

There are two muscles in the orbital group – the orbicularis oculi, which surrounds the eye socket, and the corrugator supercilii, which controls the eyebrow. The first is responsible for blinking and winking, and the second contracts to pull the eyebrows together into a frown.

We don't have a lot of control over the movement of the muscles around our nose, but the nasalis is the biggest, and with help from the depressor septi it flares the nostrils. The procerus runs from the top of the nose to the forehead, and it can pull the eyebrows down.

Finally, there are the oral muscles. The two major ones are the orbicularis oris, which surrounds the mouth and contracts to purse and pucker the lips, and the buccinator running under the cheekbone. There are also two groups of smaller muscles, the upper and the lower groups, which control the fine movements of the facial tissue to form smiles and frowns.

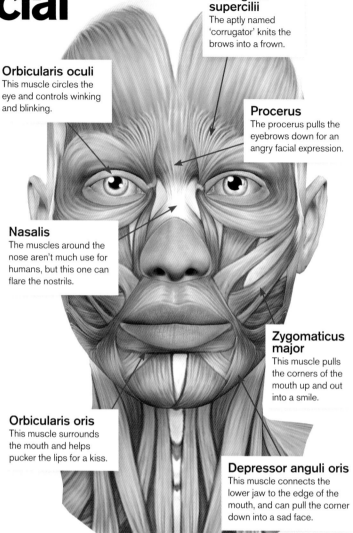

Corrugator supercilii
The aptly named 'corrugator' knits the brows into a frown.

Orbicularis oculi
This muscle circles the eye and controls winking and blinking.

Procerus
The procerus pulls the eyebrows down for an angry facial expression.

Nasalis
The muscles around the nose aren't much use for humans, but this one can flare the nostrils.

Zygomaticus major
This muscle pulls the corners of the mouth up and out into a smile.

Orbicularis oris
This muscle surrounds the mouth and helps pucker the lips for a kiss.

Depressor anguli oris
This muscle connects the lower jaw to the edge of the mouth, and can pull the corner down into a sad face.

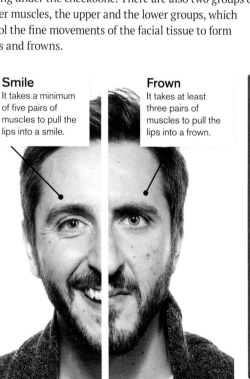

Smile
It takes a minimum of five pairs of muscles to pull the lips into a smile.

Frown
It takes at least three pairs of muscles to pull the lips into a frown.

Deciphering the face

Benjamin Amand Duchenne was a French physiologist in the 19th century, and his macabre experiments attempted to reveal the muscles responsible for different facial expressions. He wired test subjects up to galvanic probes, which delivered small electric shocks through the skin to the underlying muscles. He tried his experiment on five test subjects: a girl, a young man and woman, and an older man and woman. He captured pictures of the expressions made when different parts of the face were stimulated. Charles Darwin was so taken with the images that he later used them in his own experiments to find out whether people could read the emotions of the test subject just by looking at the expressions that their faces were pulling.

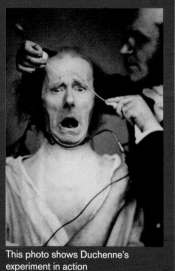

This photo shows Duchenne's experiment in action

How the spleen works

Perhaps not as well known as famous organs like the heart, the spleen serves vital functions that help keep us healthy

The spleen's main functions are to remove old blood cells and fight off infection. Red blood cells have an average life span of 120 days. Most are created from the marrow of long bones, such as the femur. When they're old, it's the spleen's job to identify them, filter them out and then break them down. The smaller particles are then sent back into the bloodstream, and either recycled or excreted from other parts of the body. This takes place in the 'red pulp', which are blood vessel-rich areas of the spleen that make up about three-quarters of its structure.

The remainder is called 'white pulp', which are areas filled with different types of immune cell (such as lymphocytes). They filter out and destroy foreign pathogens, which have invaded the body and are circulating in the blood. The white pulp breaks them down into smaller, harmless particles.

It is surrounded by a thin, fragile capsule and so is prone to injury. It sits beneath the lower ribs on the left-hand side of your body, which affords it some protection, but car crashes, major sports impacts and knife wounds can all rupture the organ. In the most serious cases, blood loss can endanger the person's life, and in these situations it needs to be removed by a surgeon. Since this reduces the body's ability to fight infections, some people will need to take antibiotics to boost their immunity for the rest of their lives.

Inside the spleen

We take you on a tour of the major features in this often-overlooked organ

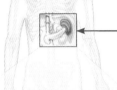

Location
The spleen sits underneath the 9th, 10th and 11th ribs (below the diaphragm) on the left-hand side of the body, which provides it with some protection against knocks.

Hilum
The entrance to the spleen, this is where the splenic artery divides into smaller branches and the splenic vein is formed from its tributaries.

Splenic artery
The spleen receives a blood supply via this artery, which arises from a branch of the aorta called the coeliac trunk.

Splenic vein
The waste products from filtration and pathogen digestion are returned to the main circulation via this vein for disposal.

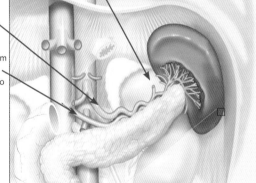

© Alamy

The immune system

Although the red blood that flows through our bodies gets all the glory, the transparent lymphatic fluid is equally important. It has its own body-wide network which follows blood vessel flow closely and allows for the transport of digested fats, immune cells and more...

Spleen
This is one of the master co-ordinators that actually staves off infections and filters old red blood cells. It contains a number of lymphocytes that recognise and destroy invading pathogens present in the blood as it flows through the spleen.

Thymus
A small organ that sits just above the heart and behind the sternum. It actually teaches T-lymphocytes to identify and destroy specific foreign bodies. Its development is directly related to hormones in the body so it's only present until puberty ends; adults don't need one.

Tonsils
These are masses of lymphoid tissue at the back of the throat and can be seen when the mouth is wide open. They form the first line of defence against inhaled foreign pathogens, although they can become infected themselves, causing tonsillitis.

Adenoids
These are part of the tonsillar system that are only present in children up until the age of five; in adults they have disappeared. They add an extra layer of defence in our early years.

Bone marrow
This forms the central, flexible part of our long bones (eg femur). Bone marrow is essential as it produces our key circulating cells, including red blood cells, white blood cells and platelets. The white blood cells mature into various types (eg lymphocytes and neutrophils), which serve as the basis of the human immune system.

Lymph nodes
These are small (about 1cm/ 0.4in) spherical nodes that are packed with macrophages and lymphocytes to defend against foreign agents. These are often linked in chains and are mainly around the head, neck, axillae (armpits) and groin.

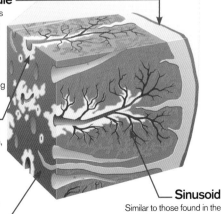

Splenic capsule
The capsule provides some protection, but it's thin and relatively weak. Strong blows or knife wounds can easily rupture it and lead to life-threatening bleeding.

White pulp
Making up roughly a quarter of the spleen, the white pulp is where white blood cells identify and destroy any type of invading pathogens.

Red pulp
Forming approximately three-quarters of the spleen, the red pulp is where red blood cells are filtered and broken down.

Sinusoid
Similar to those found in the liver, these capillaries allow for the easy passage of large cells into the splenic tissue for processing.

Kidney function

How do your kidneys filter waste from the blood to keep you alive?

K idneys are two bean-shaped organs situated halfway down the back just under the ribcage, on each side of the body, and weigh between 115 and 170 grams each, dependent on the individual's sex and size. The left kidney is commonly a little larger than the right and due to the effectiveness of these organs, individuals born with only one kidney can survive with little or no adverse health problems. Indeed, the body can operate normally with a 30-40 per cent decline in kidney function. This decline in function would rarely even be noticeable and shows just how effective the kidneys are at filtering out waste products as well as maintaining mineral levels and blood pressure throughout the body. The kidneys manage to control all of this by working with other organs and glands across the body such as the hypothalamus, which helps the kidneys determine and control water levels in the body.

Each day the kidneys will filter between a staggering 150 and 180 litres of blood, but only pass around two litres of waste down the ureters to the bladder for excretion. This waste product is primarily urea – a by-product of protein being broken down for energy – and water, and it's more commonly known as 'urine'. The kidneys filter the blood by passing it through a small filtering unit called a nephron. Each kidney has around a million of these, which are made up of a number of small blood capillaries, called glomerulus, and a urine-collecting tube called the renal tubule. The glomerulus sift the normal cells and proteins from the blood and then move the waste products into the renal tubule, which transports urine down into the bladder through the ureters.

Alongside this, the kidneys also release three hormones (known as erythropoietin, renin and calcitriol) which encourage red blood cell production, aid regulation of blood pressure and aid bone development and mineral balance respectively.

Inside your kidney

As blood enters the kidneys, it is passed through a nephron, a tiny unit made up of blood capillaries and a waste-transporting tube. These work together to filter the blood, returning clean blood to the heart and lungs for re-oxygenation and recirculation and removing waste to the bladder for excretion.

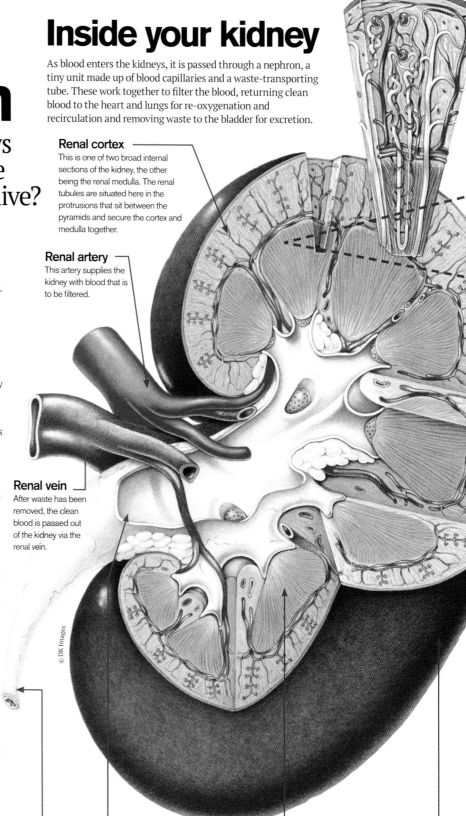

Renal cortex
This is one of two broad internal sections of the kidney, the other being the renal medulla. The renal tubules are situated here in the protrusions that sit between the pyramids and secure the cortex and medulla together.

Renal artery
This artery supplies the kidney with blood that is to be filtered.

Renal vein
After waste has been removed, the clean blood is passed out of the kidney via the renal vein.

© DK Images

Ureter
The tube that transports the waste products (urine) to the bladder following blood filtration.

Renal pelvis
This funnel-like structure is how urine travels out of the kidney and forms the top part of the ureter, which takes urine down to the bladder.

Renal medulla
The kidney's inner section, where blood is filtered after passing through numerous arterioles. It's split into sections called pyramids and each human kidney will normally have seven of these.

Renal capsule
The kidney's fibrous outer edge, which provides protection for the kidney's internal fibres.

Nephrons – the filtration units of the kidney

Nephrons are the units which filter all blood that passes through the kidneys. There are around a million in each kidney, situated in the renal medulla's pyramid structures. As well as filtering waste, nephrons regulate water and mineral salt by recirculating what is needed and excreting the rest.

Collecting duct system
Although not technically part of the nephron, this collects all waste product filtered by the nephrons and facilitates its removal from the kidneys.

Proximal tubule
Links Bowman's capsule and the loop of Henle, and will selectively reabsorb minerals from the filtrate produced by Bowman's capsule.

Glomerulus
High pressure in the glomerulus, caused by it draining into an arteriole instead of a venule, forces fluids and soluble materials out of the capillary and into Bowman's capsule.

Bowman's capsule
Also known as the glomerular capsule, this filters the fluid that has been expelled from the glomerulus. Resulting filtrate is passed along the nephron and will eventually make up urine.

Distal convoluted tubule
Partly responsible for the regulation of minerals in the blood, linking to the collecting duct system. Unwanted minerals are excreted from the nephron.

Renal artery
This artery supplies the kidney with blood. The blood travels through this, into arterioles as you travel into the kidney, until the blood reaches the glomerulus.

Renal vein
This removes blood that has been filtered from the kidney.

Loop of Henle
The loop of Henle controls the mineral and water concentration levels within the kidney to aid filtration of fluids as necessary. It also controls urine concentration.

Renal tubule
Made up of three parts, the proximal tubule, the loop of Henle and the distal convoluted tubule. They remove waste and reabsorb minerals from the filtrate passed on from Bowman's capsule.

The glomerulus

This group of capillaries is the first step of filtration and a crucial aspect of a nephron. As blood enters the kidneys via the renal artery, it is passed down through a series of arterioles which eventually lead to the glomerulus. This is unusual, as instead of draining into a venule (which would lead back to a vein) it drains back into an arteriole, which creates much higher pressure than normally seen in capillaries, which in turn forces soluble materials and fluids out of the capillaries. This process is known as ultrafiltration and is the first step in filtration of the blood. These then pass through the Bowman's capsule (also know as the glomerular capsule) for further filtration.

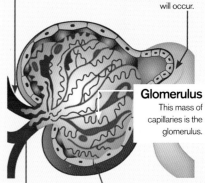

Afferent arteriole
This arteriole supplies the blood to the glomerulus for filtration.

Proximal tubule
Where reabsorption of minerals from the filtrate from Bowman's capsule will occur.

Glomerulus
This mass of capillaries is the glomerulus.

Efferent arteriole
This arteriole is how blood leaves the glomerulus following ultrafiltration.

Bowman's capsule
This is the surrounding capsule that will filter the filtrate produced by the glomerulus.

What is urine and what is it made of?

Urine is made up of a range of organic compounds such as various proteins and hormones, inorganic salts and numerous metabolites. These are often rich in nitrogen and need to be removed from the blood stream through urination. The pH-level of urine is typically around neutral (pH7) but varies depending on diet, hydration levels and physical fitness. The colour of urine is also determined by all of these different factors playing a part, with dark-yellow urine indicating dehydration and greenish urine being indicative of excessive asparagus consumption.

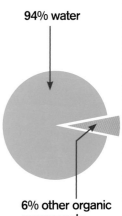

94% water

6% other organic compounds

How the liver works

The human liver is the ultimate multitasker – it performs many different functions all at the same time without you even asking

The liver is actually the largest internal organ in the human body and, has over 500 different functions. In fact, it is actually the second most complex organ after the brain and is intrinsically involved in almost every aspect of the body's metabolic processes.

The liver's main functions are energy production, removal of harmful substances and the production of crucial proteins. These tasks are carried out within liver cells, called hepatocytes, which sit in complex arrangements to maximise their overall efficiency.

The liver is the body's main powerhouse, producing and storing glucose as a key energy source. It is also responsible for breaking down complex fat molecules and building them up into cholesterol and triglycerides, which the body needs but in excess are bad. The liver makes many complex proteins, including clotting factors which are vital in arresting bleeding. Bile, which helps digest fat in the intestines, is produced in the liver and stored in the adjacent gallbladder.

The liver also plays a key role in detoxifying the blood. Waste products, toxins and drugs are

The hepatobiliary region

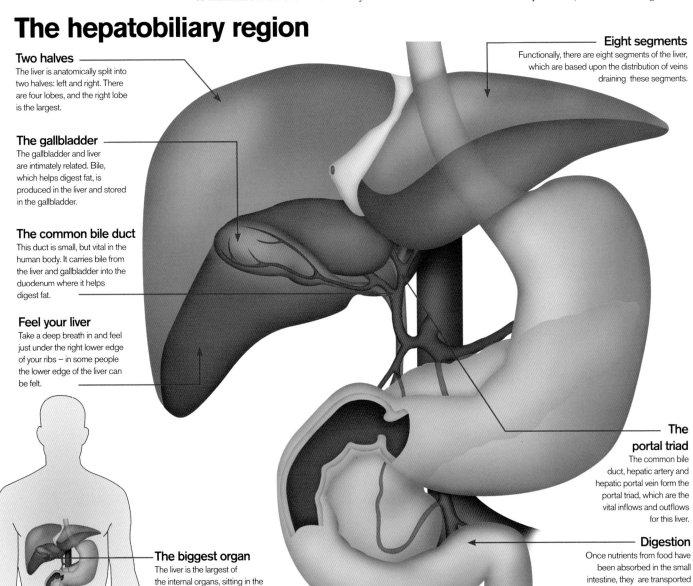

Two halves
The liver is anatomically split into two halves: left and right. There are four lobes, and the right lobe is the largest.

The gallbladder
The gallbladder and liver are intimately related. Bile, which helps digest fat, is produced in the liver and stored in the gallbladder.

The common bile duct
This duct is small, but vital in the human body. It carries bile from the liver and gallbladder into the duodenum where it helps digest fat.

Feel your liver
Take a deep breath in and feel just under the right lower edge of your ribs – in some people the lower edge of the liver can be felt.

The biggest organ
The liver is the largest of the internal organs, sitting in the right upper quadrant of the abdomen, just under the rib cage and attached to the underside of the diaphragm.

Eight segments
Functionally, there are eight segments of the liver, which are based upon the distribution of veins draining these segments.

The portal triad
The common bile duct, hepatic artery and hepatic portal vein form the portal triad, which are the vital inflows and outflows for this liver.

Digestion
Once nutrients from food have been absorbed in the small intestine, they are transported to the liver via the hepatic portal vein (not shown here) for energy production.

58

"The liver also breaks down old blood cells and recycles hormones such as adrenaline"

processed here into forms which are easier for the rest of the body to use or excrete. The liver also breaks down old blood cells, produces antibodies to fight infection and recycles hormones such as adrenaline. Numerous essential vitamins and minerals are stored in the liver: vitamins A, D, E and K, iron and copper.

Such a complex organ is also unfortunately prone to diseases. Cancers, infections (hepatitis) and cirrhosis (a form of fibrosis which is often caused by excess alcohol consumption) are just some of those which can affect the liver.

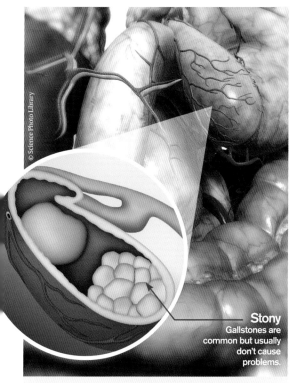

© Science Photo Library

Stony
Gallstones are common but usually don't cause problems.

The gallbladder

Bile, a dark green slimy liquid, is produced in the hepatocytes and helps to digest fat. It is stored in a reservoir which sits on the under-surface of the liver, to be used when needed. This reservoir is called the gallbladder. Stones can form in the gallbladder (gallstones) and are very common, although most don't cause problems. In 2009, just under 60,000 gallbladders were removed from patients within the NHS making it one of the most common operations performed; over 90 per cent of these are removed via keyhole surgery. Most patients do very well without their gallbladder and don't notice any changes at all.

A high demand organ

The liver deals with a massive amount of blood. It is unique because it has two blood supplies. 75 per cent of this comes directly from the intestines (via the hepatic portal vein) which carries nutrients from digestion, which the liver processes and turns into energy. The rest comes from the heart, via the hepatic artery (which branches from the aorta), carrying oxygen which the liver needs to produce this energy. The blood flows in tiny passages in between the liver cells where the many metabolic functions occur. The blood then leaves the liver via the hepatic veins to flow into the biggest vein in the body – the inferior vena cava.

Liver lobules

The functional unit which performs the liver's tasks

The liver is considered a 'chemical factory,' as it forms large complex molecules from smaller ones brought to it from the gut via the blood stream. The functional unit of the liver is the lobule – these are hexagonal-shaped structures comprising of blood vessels and sinusoids. Sinusoids are the specialised areas where blood comes into contact with the hepatocytes, where the liver's biological processes take place.

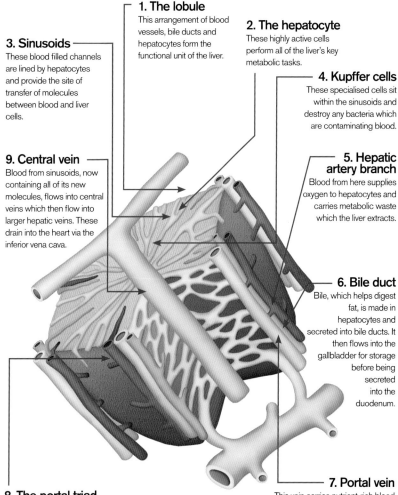

1. The lobule
This arrangement of blood vessels, bile ducts and hepatocytes form the functional unit of the liver.

2. The hepatocyte
These highly active cells perform all of the liver's key metabolic tasks.

3. Sinusoids
These blood filled channels are lined by hepatocytes and provide the site of transfer of molecules between blood and liver cells.

4. Kupffer cells
These specialised cells sit within the sinusoids and destroy any bacteria which are contaminating blood.

5. Hepatic artery branch
Blood from here supplies oxygen to hepatocytes and carries metabolic waste which the liver extracts.

9. Central vein
Blood from sinusoids, now containing all of its new molecules, flows into central veins which then flow into larger hepatic veins. These drain into the heart via the inferior vena cava.

6. Bile duct
Bile, which helps digest fat, is made in hepatocytes and secreted into bile ducts. It then flows into the gallbladder for storage before being secreted into the duodenum.

8. The portal triad
The hepatic artery, portal vein and bile duct are known as the portal triad. These sit at the edges of the liver lobule and are the main entry and exit routes for the liver.

7. Portal vein
This vein carries nutrient-rich blood directly from the intestines, which flows into sinusoids for conversion into energy within hepatocytes.

The surface area of the small intestine is huge – in fact, rolled flat it would even cover a tennis court!

Exploring the small intestine

Crucial for getting the nutrients we need from the food we eat, how does this digestive organ work?

The small intestine is actually one of the most important elements of our digestive system, which enables us to process food and absorb nutrients. On average, it sits at a little over six metres, that is 19.7 feet, long with a diameter of 2.5-3 centimetres, 1-1.2 inches. The small intestine is made up of three different distinctive parts: the duodenum, jejunum and the ileum.

The duodenum actually connects the small intestine to the stomach and is the key place for further enzyme breakdown, following already passing through the stomach, turning food into an amino acid state. While the duodenum is very important in breaking food down, using bile and enzymes from the gallbladder, liver and pancreas, it is actually the shortest element of the small bowel, only averaging about 30 centimetres, which is just 11.8 inches.

Structure of the small intestine

Examine the anatomy of this vital organ in the human digestive tract

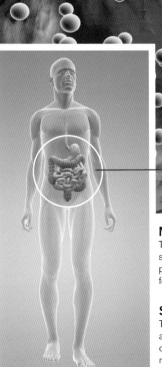

Lumen
This is the space inside the small intestine in which the food travels to be digested and absorbed.

Mucosal folds
These line the small intestine to increase surface area and help push the food on its way by creating a valve-like structure, stopping food travelling backwards.

Mucosa
The internal lining of the small intestine where the plicae circulares (mucosal folds) and villi are situated.

Submucosa
This supports the mucosa and connects it to the layers of muscle (muscularis) that make up the exterior of the small intestine.

The jejunum follows the duodenum and its primary function is to encourage absorption of carbohydrates and proteins by passing the broken-down food molecules through an area with a large surface area so they can enter the bloodstream. Villi – small finger-like structures – and mucosal folds line the passage and increase the surface area dramatically to aid this process.

The ileum is the final section of the small bowel and its main purpose is to catch nutrients that may have been missed, as well as absorbing vitamin B12 and bile salts.

Peristalsis is the movement used by the small intestine to push the food through to the large bowel, where waste matter is stored for a short period then disposed of via the colon. This process is automatically generated by a series of different muscles which make up the organ's outer wall.

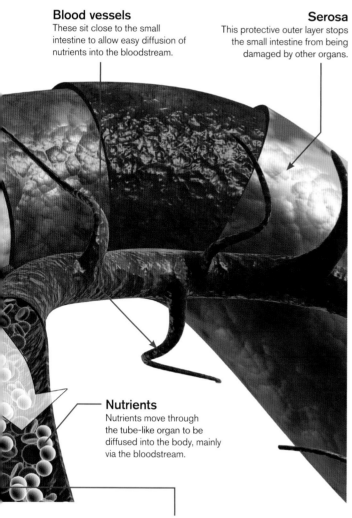

Blood vessels
These sit close to the small intestine to allow easy diffusion of nutrients into the bloodstream.

Serosa
This protective outer layer stops the small intestine from being damaged by other organs.

Nutrients
Nutrients move through the tube-like organ to be diffused into the body, mainly via the bloodstream.

What exactly are nutrients?

There are three main types of nutrient that we process in the body: lipids (fats), carbohydrates and proteins. These three groups of molecules are broken down into sugars, starches, fats and smaller, simpler molecule elements, which we can absorb through the small intestine walls and that then travel in the bloodstream to our muscles and other areas of the body that require energy or to be repaired. We also need to consume and absorb vitamins and minerals that we can't synthesise within the body, eg vitamin B12 (prevalent in meat and fish).

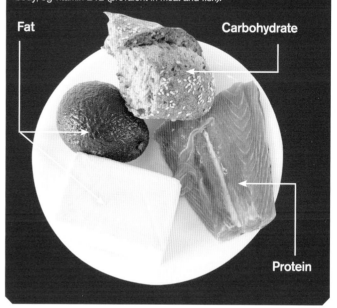

Fat

Carbohydrate

Protein

A closer look at villi

What role do these little finger-like protrusions play in the bowel?

Circular muscle layer
This works in partnership with the longitudinal muscle layer to push the food down via a process called peristalsis.

Longitudinal muscle layer
This contracts and extends to help transport food with the circular muscle layer.

Villi
Villi are tiny finger-like structures that sit all over the mucosa. They help increase the surface area massively, alongside the mucosal folds.

Epithelium (epithelial cells)
These individual cells that sit in the mucosa layer are where individual microvilli extend from.

Mucosa
The lining of the small intestine on which villi are located.

Lacteal
The lacteal is a lymphatic capillary that absorbs nutrients that can't pass directly into the bloodstream.

Microvilli
These are a mini version of villi and sit on villi's individual epithelial cells.

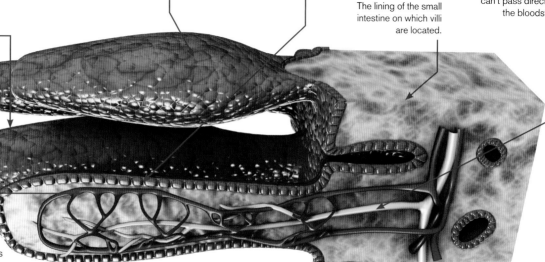

Capillary bed
These absorb simple sugars and amino acids as they pass through the epithelial tissue of the villi.

The human ribcage

Ribs are not just armour for the organs inside our torsos, as we reveal here…

The ribcage – also known as the thoracic cage or thoracic basket – is easily thought of as just a framework protecting your lungs, heart and other major organs. Although that is one key function, the ribcage does so much more. It provides vital support as part of the skeleton and, simply put, breathing wouldn't actually be possible without it.

All this means that the ribcage has to be flexible. The conical structure isn't just a rigid system of bone – it's actually both bone and cartilage. The ribcage comprises 24 ribs, joining in the back to the 12 vertebrae making up the middle of the spinal column.

The cartilage portions of the ribs meet in the front at the long, flat three-bone plate called the sternum (breastbone). Or rather, most of them do. Rib pairs one through seven are called 'true ribs' because they attach directly to the sternum. Rib pairs eight through ten attach indirectly through other cartilage structures, so they're referred to as 'false ribs'. The final two pairs – the 'floating ribs' – hang unattached to the sternum.

Rib fractures are a common and very painful injury, with the middle ribs the most likely ones to get broken. A fractured rib can be very dangerous, because a sharp piece could pierce the heart or lungs.

There's also a condition called flail chest, in which several ribs break and then detach from the cage, which can even be fatal. But otherwise there's not much you can do to mend a fractured rib other than keep it stabilised, resting and giving it time to heal.

Inside the thoracic cavity

It may not look like it at first glance, but there are more than two dozen bones that make up the ribcage…

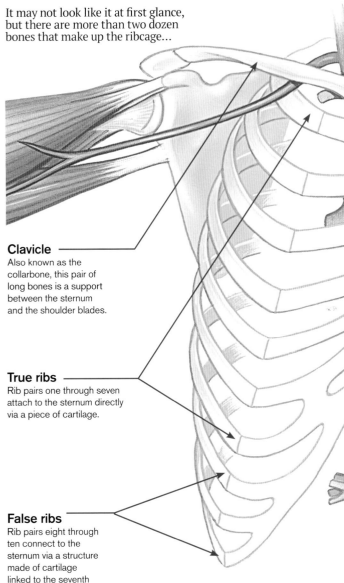

Clavicle
Also known as the collarbone, this pair of long bones is a support between the sternum and the shoulder blades.

True ribs
Rib pairs one through seven attach to the sternum directly via a piece of cartilage.

False ribs
Rib pairs eight through ten connect to the sternum via a structure made of cartilage linked to the seventh true rib.

What are hiccups?

Hiccupping – known medically as singultus, or synchronous diaphragmatic flutter (SDF) – is an involuntary spasm of the diaphragm that can happen for a number of reasons. Short-term causes include eating or drinking too quickly, a sudden change in body temperature or shock.

However, some researchers have suggested that hiccupping in premature babies – who tend to hiccup much more than full-term babies – is due to their underdeveloped lungs. It could be an evolutionary leftover, since hiccupping in humans is similar to the way that amphibians gulp water and air into their gills to breathe.

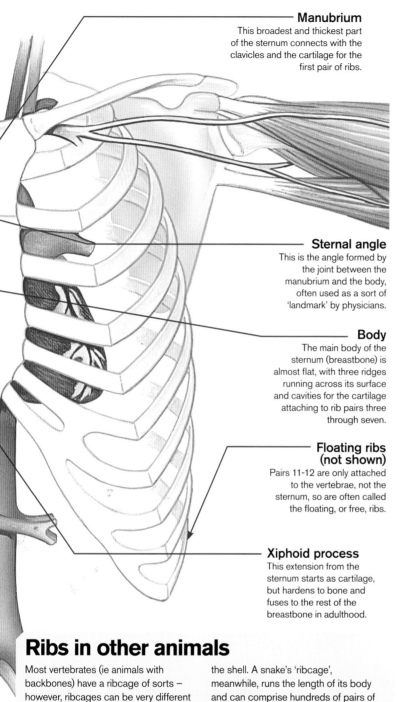

Manubrium
This broadest and thickest part of the sternum connects with the clavicles and the cartilage for the first pair of ribs.

Sternal angle
This is the angle formed by the joint between the manubrium and the body, often used as a sort of 'landmark' by physicians.

Body
The main body of the sternum (breastbone) is almost flat, with three ridges running across its surface and cavities for the cartilage attaching to rib pairs three through seven.

Floating ribs (not shown)
Pairs 11-12 are only attached to the vertebrae, not the sternum, so are often called the floating, or free, ribs.

Xiphoid process
This extension from the sternum starts as cartilage, but hardens to bone and fuses to the rest of the breastbone in adulthood.

Ribs in other animals

Most vertebrates (ie animals with backbones) have a ribcage of sorts – however, ribcages can be very different depending on the creature. For example, dogs and cats have 13 pairs of ribs as opposed to our 12. Marsupials have fewer ribs than humans, and some of those are so tiny they aren't much more than knobs of bone sticking out from the vertebrae. Once you get into other vertebrates, the differences are even greater. Birds' ribs overlap one another with hook-like structures called uncinate processes, which add strength. Frogs don't have any ribs, while turtles' eight rib pairs are fused to the shell. A snake's 'ribcage', meanwhile, runs the length of its body and can comprise hundreds of pairs of ribs. Despite the variations in appearance, ribcages all serve the same basic functions for the most part: to provide support and protection to the rest of the body.

Breathe in, breathe out…

Consciously take in a breath, and think about the fact that there are ten different muscle groups working together to make it happen. The muscles that move the ribcage itself are the intercostal muscles. They are each attached to the ribs and run between them. As you inhale, the external intercostals raise the ribs and sternum so your lungs can expand, while your diaphragm lowers and flattens. The internal intercostals lower the ribcage when you exhale. This forces the lungs to compress and release air (working in tandem with seven other muscles). If you breathe out gently, it's a passive process that doesn't require much ribcage movement.

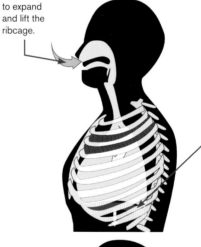

Inhalation
As you inhale, the intercostal muscles contract to expand and lift the ribcage.

Contraction
The diaphragm contracts by moving downward, allowing the lungs to fill with air.

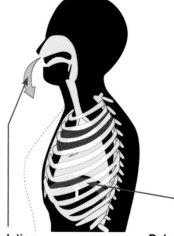

Exhalation
The intercostal muscles relax as we exhale, compressing and lowering the ribcage.

Relaxation
The diaphragm relaxes, moving upward to force air out of the lungs.

© Thinkstock

How the pancreas works

Learn how the workhorse of the digestive system helps to break down food and control our blood sugar levels

The pancreas is a pivotal organ within the digestive system. It sits inside the abdomen, behind the stomach and the large bowel, adjacent to the spleen. In humans, it has a head, neck, body and tail. It is connected to the first section of the small intestine, the duodenum, by the pancreatic duct, and to the bloodstream via a rich network of vessels. When it comes to the function of the pancreas, it is best to think about the two types of cell it contains: endocrine and exocrine.

The endocrine pancreas is made up of clusters of cells called islets of Langerhans, which in total contain approximately 1 million cells and are responsible for producing hormones. These cells include alpha cells, which secrete glucagon, and beta cells which generate insulin. These two hormones have opposite effects on blood sugar levels throughout the body: glucagon increases glucose levels, while insulin decreases them.

The cells here are all in contact with capillaries, so hormones which are produced can be fed directly into the bloodstream. Insulin secretion is under the control of a negative-feedback loop; high blood sugar will lead to insulin secretion, which then lowers blood sugar with subsequent suppression of insulin. Disorders of these cells (and thus alterations of the hormone levels) can lead to many serious conditions, including diabetes. The islets of Langerhans are also responsible for producing other hormones, like somatostatin, which governs nutrient absorption among many other things.

The exocrine pancreas, meanwhile, is responsible for secreting digestive enzymes. Cells are arranged in clusters called acini, which flow into the central pancreatic duct. This leads into the duodenum – part of the small bowel – to come into contact with and aid in the digestion of food. The enzymes secreted include proteases (to digest protein), lipases (for fat) and amylase (for sugar/starch). Secretion of these enzymes is controlled by a series of hormones, which are released from the stomach and duodenum in response to the stretch from the presence of food.

Anatomy of the pancreas

It might not be the biggest organ but the pancreas is a key facilitator of how we absorb nutrients and stay energised

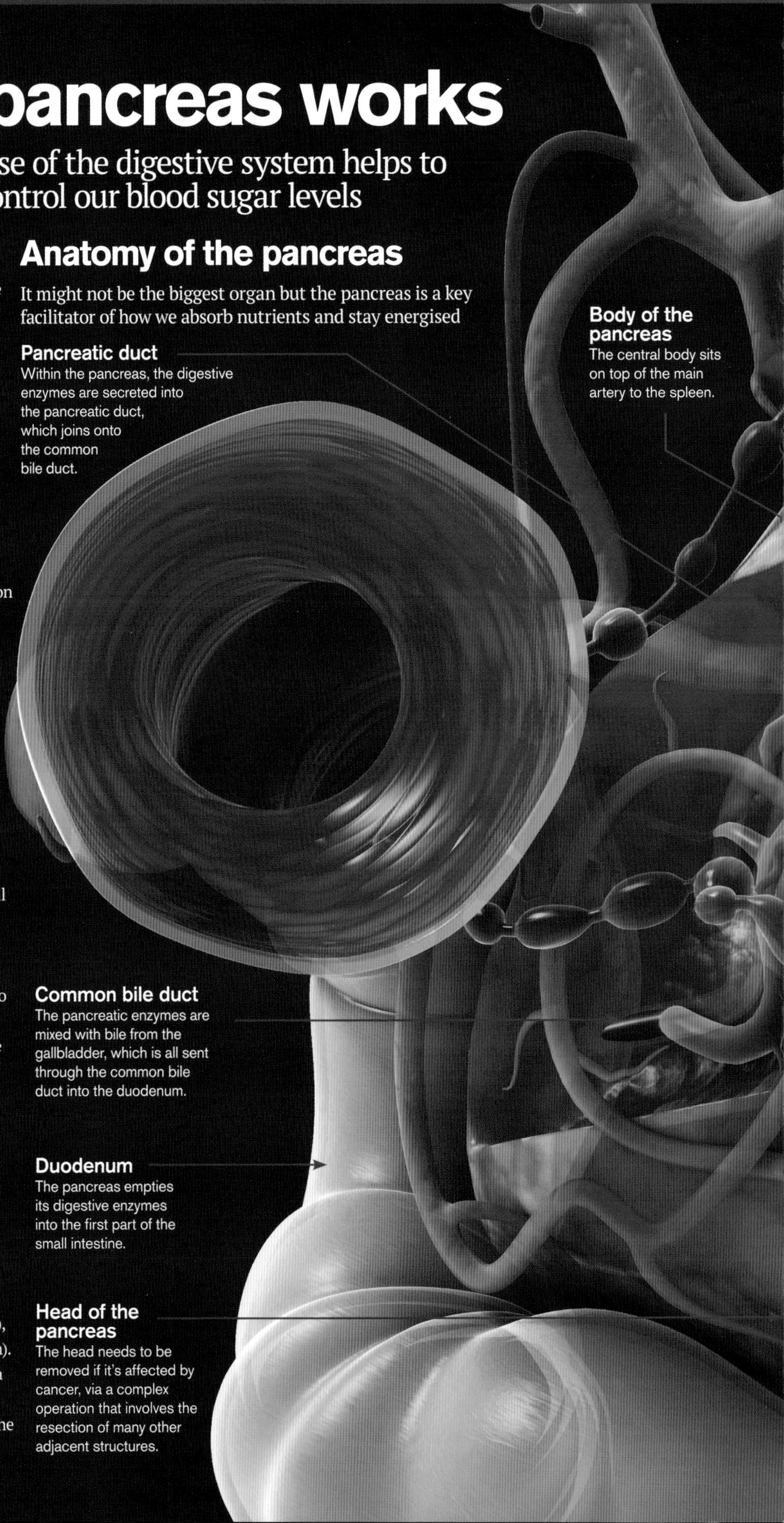

Pancreatic duct
Within the pancreas, the digestive enzymes are secreted into the pancreatic duct, which joins onto the common bile duct.

Body of the pancreas
The central body sits on top of the main artery to the spleen.

Common bile duct
The pancreatic enzymes are mixed with bile from the gallbladder, which is all sent through the common bile duct into the duodenum.

Duodenum
The pancreas empties its digestive enzymes into the first part of the small intestine.

Head of the pancreas
The head needs to be removed if it's affected by cancer, via a complex operation that involves the resection of many other adjacent structures.

Tail of the pancreas
This is the end portion of the organ and is positioned close to the spleen.

What brings on diabetes?

Diabetes is a condition where a person has higher blood sugar than normal. It is either caused by a failure of the pancreas to produce insulin (ie type 1, or insulin-dependent diabetes mellitus), or resistance of the body's cells to insulin present in the circulation (ie type 2, or non-insulin-dependent diabetes mellitus). There are also other disorders of the

pancreas. Inflammation of the organ (ie acute pancreatitis) causes severe pain in the upper abdomen, forcing most people to attend the emergency department as it can actually be life threatening. In contrast, cancer of the pancreas causes the individual gradually worsening pain which can commonly be mistaken for various other ailments.

Beta cells
It is the beta cells within the islets of Langerhans which control glucose levels and amount of insulin secretion.

Insulin released
The vesicle releases its stored insulin into the blood capillaries through exocytosis.

High glucose
When the levels of glucose within the bloodstream are high, the glucose wants to move down its diffusion gradient into the cells.

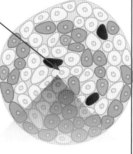

Calcium effects
The calcium causes the vesicles that store insulin to move towards the cell wall.

Blood supply
The pancreas derives its blood supply from a variety of sources, including vessels running to the stomach and spleen.

Does the pancreas vary in humans and animals?

Every vertebrate animal has a pancreas of some form, meaning they are all susceptible to diabetes too. The arrangement, however, varies from creature to creature. In humans, the pancreas is most often a single structure that sits at the back of the abdomen. In other animals, the arrangement varies from two or three masses of tissue scattered around the abdomen, to tissue interspersed within the connective tissue between the bowels, to small collections of tissue within the bowel mucosal wall itself. One of the other key differences is the number of ducts that connect the pancreas to the bowel. In most humans there's only one duct, but occasionally there may be two or three – and sometimes even more. In other animals, the number is much more variable. However, the function is largely similar, where the pancreas secretes digestive enzymes and hormones to control blood sugar levels.

GLUT2
This is a glucose-transporting channel, which facilitates the uptake of glucose into the cells.

Depolarisation
The metabolism of glucose leads to changes in the polarity of the cell wall and an increase in the number of potassium ions.

Calcium channels
Changes in potassium levels cause voltage-gated calcium channels to open in the cell wall, and calcium ions to flow into the cell.

© Corbis; Süleyman Habib

65

The urinary system explained

Every day the body produces waste products that enter the bloodstream – but how do we get rid of them?

The human urinary system's primary function is to remove by-products which remain in the blood after the body has metabolised food. The process is made up of several different key features. Generally, this system consists of two kidneys, two ureters, the bladder, two sphincter muscles (one internal, one external) and a urethra and these work alongside the intestines, lungs and skin, all of which excrete waste products from the body.

The abdominal aorta is an important artery to the system as this feeds the renal artery and vein, which supply the kidneys with blood. This blood is filtered by the kidneys to remove waste products, such as urea which is formed through amino acid metabolism. Through communication with other areas of the body, such as the hypothalamus, the kidneys also control water levels in the body, sodium and potassium levels among other electrolytes, blood pressure, pH of the blood and are also involved in red blood cell production through the creation and release of the hormone erythropoietin. Consequently, they are absolutely crucial to optimum body operation.

After blood has been filtered by the kidneys, the waste products then travel down the ureters to the bladder. The bladder's walls expand out to hold the urine until the body can excrete the waste out through the urethra. The internal and external sphincters then control the release of urine.

On average, a typical human will produce approximately a staggering 2.5-3 litres of urine in just one day, although this can vary dramatically dependant on external factors such as how much water is consumed.

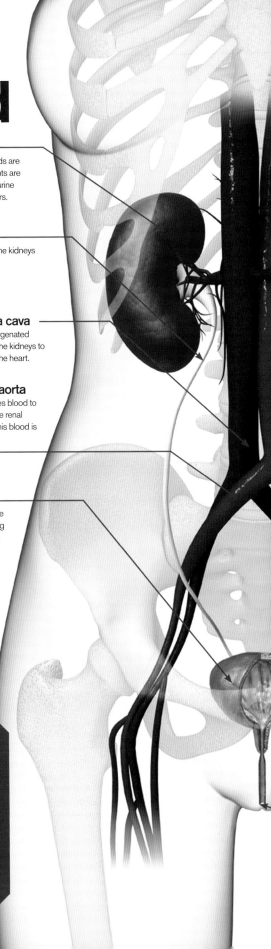

Kidneys
This is where liquids are filtered and nutrients are absorbed before urine exits into the ureters.

Ureter
These tubes link the kidneys and the bladder.

Inferior vena cava
This carries deoxygenated blood back from the kidneys to the right aorta of the heart.

Abdominal aorta
This artery supplies blood to the kidneys, via the renal artery and vein. This blood is then cleansed by the kidneys.

Bladder
This is where urine gathers after being passed down the ureters from the kidneys.

How do the kidneys work?

The kidneys will have around 150-180 litres of blood to filter per day, but only pass around two litres of waste down the ureters to the bladder for excretion, therefore the kidneys return much of this blood, minus most of the waste products, to the heart for re-oxygenation and recirculation around the body.

The way the kidneys do this is to pass the blood through a small filtering unit called a nephron. Each kidney has around a million of these, which are made up of a number of small blood capillaries and a tube called the renal tubule. The blood capillaries sift the normal cells and proteins from the blood for recirculation and then direct the waste products into the renal tubule. This waste, which will primarily consist of urea, mixes with water and forms urine as it passes through the renal tubule and then into the ureter on its way to the bladder.

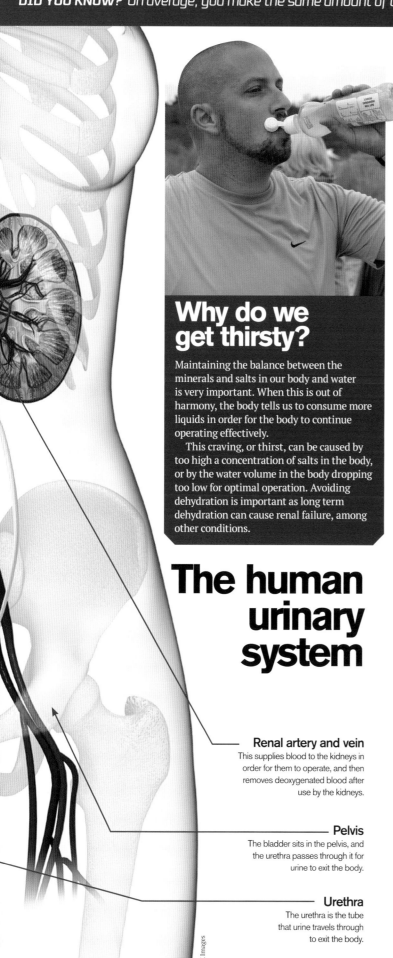

How do we store waste until we're ready to expel it?

The bladder stores waste products by allowing the urine to enter through the ureter valves, which attach the ureter to the bladder. The walls relax as urine enters and this allows the bladder to stretch. When the bladder becomes full, the nerves in the bladder communicate with the brain and cause the individual to feel the urge to urinate. The internal and external sphincters will then relax, allowing urine to pass down the urethra.

Bladder fills

1. Ureters
These tubes connect to the kidneys and urine flows down to the bladder through them.

2. Internal urethral sphincter
This remains closed to ensure urine does not escape unexpectedly.

3. External urethral sphincter
This secondary sphincter also remains closed to ensure no urine escapes.

4. Ureter valves
These valves are situated at the end of the ureters and let urine in.

5. Bladder walls (controlled by detrusor muscles)
The detrusor muscles in the wall of the bladder relax to allow expansion of the bladder as necessary.

Bladder empties

1. Internal urethral sphincter
This relaxes when the body is ready to expel the waste.

2. External urethral sphincter
This also relaxes for the urine to exit the body.

3. Bladder walls (controlled by detrusor muscles)
These muscles contract to force the urine out of the bladder.

4. Urethra
Urine travels down this passageway to exit the body.

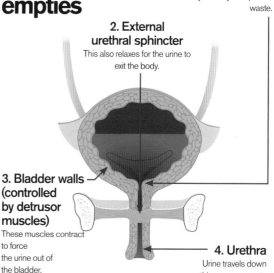

Why do we get thirsty?

Maintaining the balance between the minerals and salts in our body and water is very important. When this is out of harmony, the body tells us to consume more liquids in order for the body to continue operating effectively.

This craving, or thirst, can be caused by too high a concentration of salts in the body, or by the water volume in the body dropping too low for optimal operation. Avoiding dehydration is important as long term dehydration can cause renal failure, among other conditions.

The human urinary system

Renal artery and vein
This supplies blood to the kidneys in order for them to operate, and then removes deoxygenated blood after use by the kidneys.

Pelvis
The bladder sits in the pelvis, and the urethra passes through it for urine to exit the body.

Urethra
The urethra is the tube that urine travels through to exit the body.

© DK Images

Inside the human stomach

Discover how this amazing digestive organ stretches, churns and holds corrosive acid to break down our food, all without getting damaged

The stomach's major role is as a reservoir for food; it allows large meals to be consumed in one sitting before being gradually emptied into the small intestine. A combination of acid, protein-digesting enzymes and vigorous churning action breaks the stomach contents down into an easier-to-process liquid form, preparing food for absorption in the bowels.

In its resting state, the stomach is contracted and the internal surface of the organ folds into characteristic ridges, or rugae. When we start eating, however, the stomach begins to distend; the rugae flatten, allowing the stomach to expand, and the outer muscles relax. The stomach can accommodate about a litre (1.8 pints) of food without discomfort.

The expansion of the stomach activates stretch receptors, which trigger nerve signalling that results in increased acid production and powerful muscle contractions to mix and churn the contents. Gastric acid causes proteins in the food to unravel, allowing access by the enzyme pepsin, which breaks down protein. The presence of partially digested proteins stimulates enteroendocrine cells (G-cells) to make the hormone gastrin, which encourages even more acid production.

The stomach empties its contents into the small intestine through the pyloric sphincter. Liquids pass through the sphincter easily, but solids must be smaller than one to two millimetres (0.04-0.08 inches) in diameter before they will fit. Anything larger is 'refluxed' backwards into the main chamber for further churning and enzymatic breakdown. It takes about two hours for half a meal to pass into the small intestine and the process is generally complete within four to five hours.

Lining under the microscope

The stomach is much more than just a storage bag. Take a look at its complex microanatomy now…

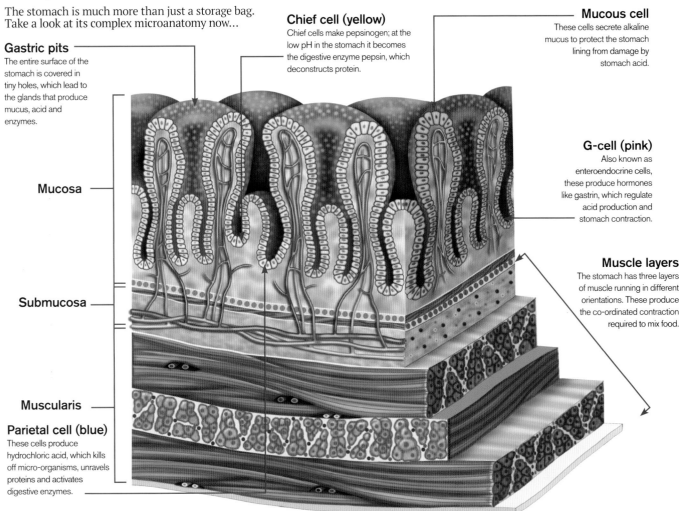

Gastric pits
The entire surface of the stomach is covered in tiny holes, which lead to the glands that produce mucus, acid and enzymes.

Chief cell (yellow)
Chief cells make pepsinogen; at the low pH in the stomach it becomes the digestive enzyme pepsin, which deconstructs protein.

Mucous cell
These cells secrete alkaline mucus to protect the stomach lining from damage by stomach acid.

Mucosa

G-cell (pink)
Also known as enteroendocrine cells, these produce hormones like gastrin, which regulate acid production and stomach contraction.

Submucosa

Muscle layers
The stomach has three layers of muscle running in different orientations. These produce the co-ordinated contraction required to mix food.

Muscularis

Parietal cell (blue)
These cells produce hydrochloric acid, which kills off micro-organisms, unravels proteins and activates digestive enzymes.

Gastric anatomy

This major organ in the digestive system has several distinct regions with different functions, as we highlight here

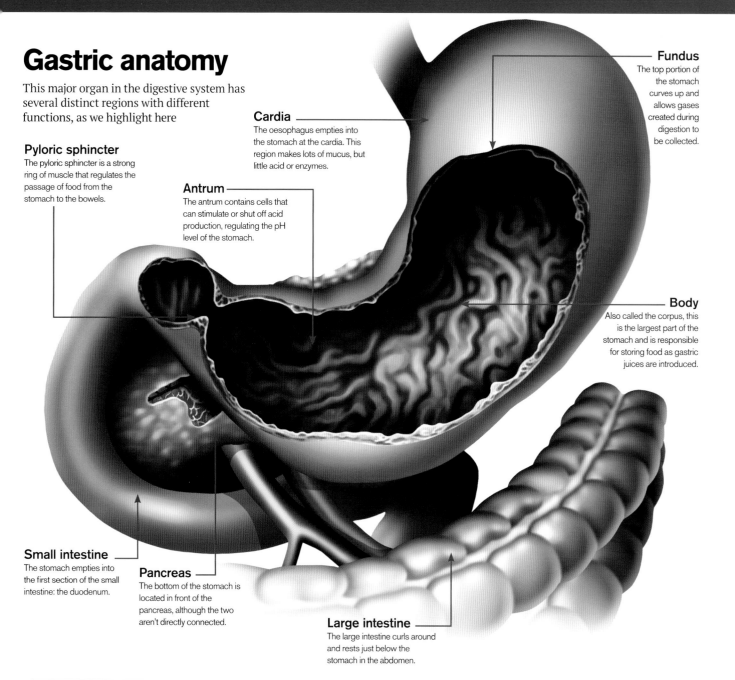

Fundus
The top portion of the stomach curves up and allows gases created during digestion to be collected.

Cardia
The oesophagus empties into the stomach at the cardia. This region makes lots of mucus, but little acid or enzymes.

Pyloric sphincter
The pyloric sphincter is a strong ring of muscle that regulates the passage of food from the stomach to the bowels.

Antrum
The antrum contains cells that can stimulate or shut off acid production, regulating the pH level of the stomach.

Body
Also called the corpus, this is the largest part of the stomach and is responsible for storing food as gastric juices are introduced.

Small intestine
The stomach empties into the first section of the small intestine: the duodenum.

Pancreas
The bottom of the stomach is located in front of the pancreas, although the two aren't directly connected.

Large intestine
The large intestine curls around and rests just below the stomach in the abdomen.

Why doesn't it digest itself?

Your stomach is full of corrosive acid and enzymes capable of breaking down protein – left unprotected the stomach lining would quic be destroyed. To prevent this from occurring, cells lining the stomach wall produce carbohydrate-rich mucus, which forms a slipp gel-like barrier. The mucus contains bicarbona which is alkaline and buffers the pH at the surface of the stomach lining, preventing damage by acid. For added protection, the protein-digesting enzyme pepsin is created fr a zymogen (the enzyme in its inactive form) – pepsinogen; it only becomes active when it comes into contact with acid, a safe distance away from the cells that manufacture it.

Produced by parietal cells in the stomach lining, gastric acid has a pH level of 1.5 to 3.5

Vomit reflex step-by-step

Vomiting is the forceful expulsion of the stomach contents up the oesophagus and out of the mouth. It's the result of three co-ordinated stages. First, a deep breath is drawn and the body closes the glottis, covering the entrance to the lungs. The diaphragm then contracts, lowering pressure in the thorax to open up the oesophagus. At the same time, the muscles of the abdominal wall contract, which squeezes the stomach. The combined shifts in pressure both inside and outside the stomach forces any contents upwards.

Bones in the hand

The human hand contains 27 bones, and these divide up into three distinct groups: the carpals, metacarpals and phalanges. These also then break down into a further three different groups: the proximal phalanges, the intermediate phalanges and then the distal phalanges. Eight bones are situated in the wrist and these are collectively called the carpals. The metacarpals, which are situated in the palm of the hand account for a further five out of the 27, and each finger has three phalanges, the thumb only has two. Intrinsic muscles and tendons control movement of the digits and hand, and attach to extrinsic muscles that extend further up into the arm, flexing the digits.

The human hand

We take our hands for granted, but they are actually quite complex and have been crucial in our evolution

The human hand is an important feature of the human body, which allows individuals to manipulate their surroundings and also to gather large amounts of data from the environment that the individual is situated within. A hand is generally defined as the terminal aspect of the human arm, which consists of prehensile digits, an opposable thumb, and a wrist and palm. Although many other animals have similar structures, only primates and a limited number of other vertebrates can be said to have a 'hand' due to the need for an opposable thumb to be present and the degree of extra articulation that the human hand can achieve. Due to this extra articulation, humans have developed fine motor skills allowing for much increased control in this limb. Consequently we see improved ability to grasp and grip items and development of skills such as writing.

A normal human hand is made up of five digits, the palm and wrist. It consists of 27 bones, tendons, muscles and nerves, with each fingertip of each digit containing numerous nerve endings making the hand a crucial area for gathering information from the environment using one of man's most crucial five senses: touch. The muscles interact together with tendons in order to allow fingers to bend, straighten, point and, in the case of the thumb, rotate. However, the hand is an area that sees many injuries due to the number of ways we use it, one in ten injuries in A&E being hand related, and there are also several disorders that can affect the hand development whilst still in the womb, such as polydactyly, where an individual is born with extra digits, which are often still in perfect working order.

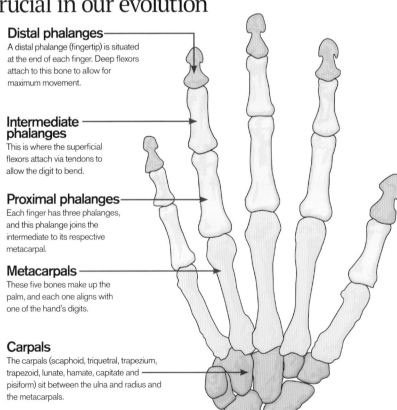

Distal phalanges
A distal phalange (fingertip) is situated at the end of each finger. Deep flexors attach to this bone to allow for maximum movement.

Intermediate phalanges
This is where the superficial flexors attach via tendons to allow the digit to bend.

Proximal phalanges
Each finger has three phalanges, and this phalange joins the intermediate to its respective metacarpal.

Metacarpals
These five bones make up the palm, and each one aligns with one of the hand's digits.

Carpals
The carpals (scaphoid, triquetral, trapezium, trapezoid, lunate, hamate, capitate and pisiform) sit between the ulna and radius and the metacarpals.

Muscles and other structures

The movements and articulations of the hand and by the digits are not only controlled by tendons but also two muscle groups situated within the hand and wrist. These are the extrinsic and intrinsic muscle groups, so named as the extrinsics are attached to muscles which extend into the forearm, whereas the intrinsics are situated within the hand and wrist. The flexors and extensors, which make up the extrinsic muscles, use either exclusively tendons to attach to digits they control (flexors)

or a more complex mix of tendons and intrinsic muscles to operate (extensors). These muscles will contract in order to cause digit movement, and flexors and extensors work in a pair to complement each to straighten and bend digits. The intrinsic muscles are responsible for aiding all extrinsic muscle action and any other movements in the digits and have three distinct groups; the thenar and hypothenar (referring to the thumb and little finger respectively), the interossei and the lumbrical.

Opposable thumbs

Increased articulation of the thumb has been heralded as one of the key factors in human evolution. It allowed for increased control and grip, and has allowed for tool use in order to develop among human ancestors as well as other primates. This has later also facilitated major cultural advances, such as writing. Alongside the four other flexible digits, the opposable thumb makes the human hand one of the most dexterous in the world. A thumb can only be classified as opposable when it can be brought opposite to the other digits.

Left handed or right handed?

The most common theory for why some individuals are left handed is that of the 'disappearing twin'. This supposes that the left-handed individual was actually one of a set of twins, but that in the early stages of development the other, right handed, twin died. However, it's been found that dominance of one hand is directly linked with hemisphere dominance in the brain, as in many other paired organs.

Individuals who somehow damage their dominant hand for extended periods of time can actually change to use the other hand, proving the impact and importance of environment and extent to which humans can adapt.

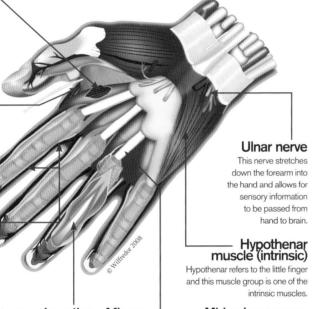

Thenar space
Thenar refers to the thumb, and this space is situated between the first digit and thumb. One of the deep flexors (extrinsic muscle) is located in here.

Interossei muscle (intrinsic)
This interossei muscle sits between metacarpal bones and will unite with tendons to allow extension using extrinsic muscles.

Arteries, veins and nerves
These supply fresh oxygenated blood (and take away deoxygenated blood) to hand muscles.

© Wilfredor 2008

Ulnar nerve
This nerve stretches down the forearm into the hand and allows for sensory information to be passed from hand to brain.

Hypothenar muscle (intrinsic)
Hypothenar refers to the little finger and this muscle group is one of the intrinsic muscles.

Insertion of flexor tendon
This is where the tendon attaches the flexor muscle to the finger bones to allow articulation.

Mid palmar space
Tendons and intrinsic muscles primarily inhabit this space within the hand.

Forearm muscles

Extrinsic muscles are so called because they are primarily situated outside the hand, the body of the muscles situated along the underside or front of the forearm. This body of muscles actually breaks down into two quite distinct groups: the flexors and the extensors. The flexors run alongside the underside of the arm and are responsible for allowing the bending of the individual digits, whereas the extensor muscles' main purpose is the reverse this action, to straighten the digits. There are both deep and superficial flexors and extensors, and which are used at any one time depends on the digit to be moved.

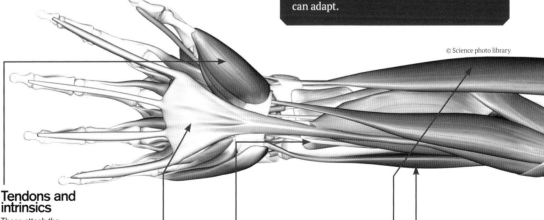

© Science photo library

Tendons and intrinsics
These attach the flexor muscles to the phalanges, and facilitate bending. Tendons also interact with the intrinsics and extensors in the wrist, palm and forearm to straighten the digits.

Thenars
The intrinsic group of muscles is used to flex the thumb and control its sideways movement.

Superficial flexors
The other flexor that acts on the digits is the superior flexor, which attaches to the intermediate phalanges.

Deep flexors
The digits have two extrinsic flexors that allow them to bend, the deep flexor and the superficial. The deep flexor attaches to the distal phalanges.

Extensors
Extensors on the back of the forearm straighten the digits. Divided into six sections, their connection to the digits is complex.

How do your feet work?

Feet are immensely complex structures, yet we put huge amounts of pressure on them every day. How do they cope?

The human foot and ankle is crucial for locomotion and is one of the most complex structures of the human body. This intricate structure is made up of no less than 26 bones, 20 muscles, 33 joints – although only 20 are articulated – as well as numerous tendons and ligaments. Tendons connect the muscles to the bones and facilitate movement of the foot, while ligaments hold the tendons in place and help the foot move up and down to initiate walking. Arches in the foot are formed by ligaments, muscles and foot bones and help to distribute weight, as well as making it easier for the foot to operate efficiently when walking and running. It is due to the unique structure of the foot and the way it distributes pressure throughout all aspects that it can withstand constant pressure throughout the day.

One of the other crucial functions of the foot is to aid balance, and toes are a crucial aspect of this. The big toe in particular helps in this area, as we can grip the ground with it if we feel we are losing balance.

The skin, nerves and blood vessels make up the rest of the foot, helping to hold the shape and also supplying it with all the necessary minerals, oxygen and energy to help keep it moving easily and constantly.

What happens when you sprain your ankle?

A sprained ankle is the most common type of soft tissue injury. The severity of the sprain can depend on how you sprained the ankle, and a minor sprain will generally consist of a stretched or only partially torn ligament. However, more severe sprains can cause the ligament to tear completely, or even force a piece of bone to break off.

Generally a sprain will happen when you lose balance or slip, and the foot bends inwards towards the other leg. This then overstretches the ligaments and causes the damage. Actually, over a quarter of all sporting injuries are sprains of the ankle.

© DK Images

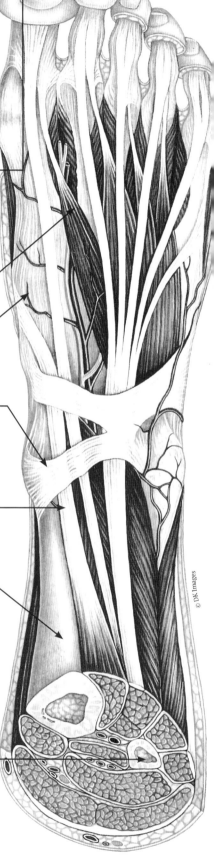

Toes
Terminal aspects of the foot that aid balance by grasping onto the ground. They are the equivalent of fingers in the foot structure.

Muscles – including the extensor digitorum brevis muscle
Muscles within the foot help the foot lift and articulate as necessary. The extensor digitorum brevis muscle sits on the top of the foot, and helps flex digits two-four on the foot.

Blood vessels
These supply blood to the foot, facilitating muscle operation by supplying energy and oxygen and removing deoxygenated blood.

Ligaments
Ligaments support the tendons and help to form the arches of the foot, spreading weight across it.

Tendons (extensor digitorum longus, among others)
Fibrous bands of tissue which connect muscles to bones. They can withstand a lot of tension and link various aspects of the foot, facilitating movement.

Tibia
The larger and stronger of the lower leg bones, this links the knee and the ankle bones of the foot.

Fibula
This bone sits alongside the tibia, also linking the knee and the ankle.

© DK Images

The structure of the foot and how the elements work together

How do we walk?

'Human gait' is the term to describe how we walk. This gait will vary between each person, but the basics are the same

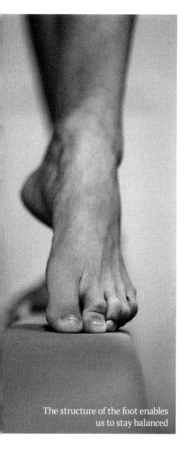

The structure of the foot enables us to stay balanced

2. Weight transfer
The weight will transfer fully to the foot still in contact with the ground, normally with a slight leaning movement of the body.

4. Leg swing
The lower leg will then swing at the knee, under the body, to be placed in front of the stationary, weight-bearing foot.

3. Foot lift
After weight has transferred and the individual feels balanced, the ball of the first foot will then lift off the ground, raising the thigh.

5. Heel placement
The heel will normally be the part of the foot that's placed first, and weight will start to transfer back onto this foot as it hits the ground.

1. Heel lift
The first step of walking is for the foot to be lifted off the ground. The knee will raise and the calf muscle and Achilles tendon, situated on the back of the leg, will contract to allow the heel to lift off the ground.

6. Repeat process
The process is then repeated with the other foot. During normal walking or running, one foot will start to lift as the other starts to come into contact with the ground.

Bones of the foot

Distal phalanges
The bones which sit at the far end of the foot and make up the tips of the toes.

Proximal phalanges
These bones link the metatarsals and the distal phalanges and stretch from the base of the toes.

Metatarsals
The five, long bones that are the metatarsals are located between the tarsal bones and the phalanges. These are the equivalent of the metacarpals in the hand.

Cuneiforms bones (three)
Three bones that fuse together during bone development and sit between the metatarsals and the talus.

Navicular
This bone, which is so named due to its resemblance to a boat, articulates with the three cuneiform bones.

A baby is born with 22 out of a total 26 bones in each foot

Cuboid
One of five irregular bones (cuboid, navicular and three cuneiform bones) which make up the arches of the foot. These help with shock absorption in locomotion.

Talus
The talus is the second largest bone of the foot, and it makes up the lower part of the ankle joint.

Calcaneus
This bone constitutes the heel and is crucial for walking. It is the largest bone in the foot.

© DK Images

73

HACKING THE HUMAN BODY

YOUR BODY IS YOUR MOST VERSATILE TOOL, BUT WHAT IF YOU COULD IMPROVE IT?

We are limited by our biology: prone to illness, doomed to wear out over time, and restricted to the senses and abilities that nature has crafted for us over millions of years of evolution. But not any more.

Biological techniques are getting cheaper and more powerful, electronics are getting smaller, and our understanding of the human body is growing. Pacemakers already keep our hearts beating, hormonal implants control our fertility, and smart glasses augment our vision. We are teetering on the edge of the era of humanity 2.0, and some enterprising individuals have already made the leap to the other side.

While much of the technology developed so far has had a medical application, people are now choosing to augment their healthy bodies to extend and enhance their natural abilities.

Kevin Warwick, a professor of cybernetics at Coventry University, claims to be the "world's first cyborg". In 1998, he had a silicon chip implanted into his arm, which allowed him to open doors, turn on lights and activate computers without even touching them. In 2002, the system was upgraded to communicate with his nervous system; 100 electrodes were linked up to his median nerve.

Through this new implant, he could control a wheelchair, move a bionic arm and, with the help of a matched implant fitted into his wife, he was even able to receive nerve impulses from another human being.

Professor Warwick's augmentations were the product of a biomedical research project, but waiting for these kinds of modifications to hit the mainstream is proving too much for some enterprising individuals, and hobbyists are starting to experiment for themselves.

Amal Graafstra is based in the US, and is a double implantee. He has a Radio Frequency Identification (RFID) chip embedded in each hand: the left opens his front door and starts his motorbike, and the right stores data uploaded from his mobile phone. Others have had magnets fitted inside their fingers, allowing them to sense magnetic fields, and some are experimenting with aesthetic implants, putting silicon shapes and lights beneath their skin. Meanwhile, researchers are busy developing the next generation of high-tech equipment to upgrade the body still further.

This article comes with a health warning: we don't want you to try this at home. But it's an exciting glimpse into some of the emerging technology that could be used to augment our bodies in the future. Let's dive in to the sometimes shady world of biohacking.

Implants

Professional and amateur biohackers are exploring different ways of augmenting our skin

Electronic tattoos

Not so much an implant as a stick-on mod, this high-tech tattoo from the Massachusetts Institute of Technology (MIT) can store information, change colour, and even control your phone.

Created by the MIT Media Lab and Microsoft Research, DuoSkin is a step forward from the micro-devices that fit in clothes, watches and other wearables. These tattoos use gold leaf to conduct electricity against the skin, performing three main functions: input, output and communication.

Some of the electronic tattoos work similarly to buttons or touch pads. Others change colour using resistors and temperature-sensitive chemicals, and some contain coils that can be used for wireless communication.

The electronic tattoos work as touch sensors, change colour, and receive Wi-Fi signals

Fingertip magnets

Tiny neodymium magnets can be coated in silicon and implanted into the fingertips. They respond to magnetic fields produced by electrical wires, whirring fans and other tech. This gives the wearer a 'sixth sense', allowing them to pick up on the shape and strength of invisible fields in the air.

The implants allow the wearer to pick up small magnetic objects

Under-skin lights

Some implants are inserted under the skin to augment the appearance of the body. The procedure involves cutting and stitching, and is often performed by tattoo artists or body piercers. The latest version, created by a group in Pittsburgh, even contains LED lights. This isn't for the faint of heart – anaesthetics require a license, so fitting these is usually done without.

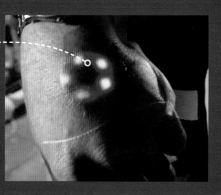

Grindhouse Wetware makes implantable lights that glow from under the skin

Buzzing the brain

Transcranial DC stimulation sends electrical signals through the skull to enhance performance

Motor control
If the current is applied over the motor cortex, it increases excitability of the nerve cells responsible for movement.

Visual perception
Visual information is processed at the back of the brain, and electrodes placed here can augment our ability to interpret our surroundings.

Excitability
The electricity changes the activity of the nerve cells in the brain, making them more likely to fire.

Working memory
Stimulation of the front of the brain seems to improve short-term memory and learning.

Wires
A weak current of around one to two milliamperes is delivered to the brain for 10 to 30 minutes.

Cathode
Current moves towards the cathode completing the circuit. Changing the placement of the electrodes alters the effect on brain function.

Device
Powered by a simple nine-volt battery, the device delivers a constant current to the scalp.

Anode
The anode delivers current from the device across the scalp and into the brain.

Gene editing

In 2013, researchers working in gene editing made a breakthrough. They used a new technique to cut the human genome at sites of their choosing, opening the floodgates for customising and modifying our genetics.

The system that they used is called CRISPR. It is adapted from a system found naturally in bacteria, and is composed of two parts: a Cas9 enzyme that acts like a pair of molecular scissors, and a guide molecule that takes the scissors to a specific section of DNA.

What scientists have done more recently is to hijack this system. By 'breaking' the enzyme scissors, the CRISPR system no longer cuts the DNA. Instead, it can be used to switch the genes on and off at will, without changing the DNA sequence. At the moment, the technique is still experimental, but in the future it could be used to repair or alter our genes.

The CRISPR complex works like a pair of DNA-snipping scissors

Hacking the brain

With the latest technology we can decipher what the brain is thinking, and we can talk back

The human brain is the most complex structure in the known universe, but ultimately it communicates using electrical signals, and the latest tech can tap into these coded messages.

Prosthetic limbs can now be controlled by the mind; some use implants attached to the surface of the brain, while others use caps to detect electrical activity passing across the scalp. Decoding signals requires a lot of training, and it's not perfect, but year after year it is improving.

It is also possible to communicate in the other direction, sending electrical signals into the brain. Retinal implants can pick up light, code it into electrical pulses and deliver them to the optic nerve, and cochlear implants do the same with sound in the ears via the cochlear nerve. And, by attaching electrodes to the scalp, whole areas of the brain can be tweaked from the outside.

Transcranial direct current stimulation uses weak currents that pass through skin and bone to the underlying brain cells. Though still in development, early tests indicate that this can have positive effects on mood, memory and other brain functions. The technology is relatively simple, and companies are already offering the kit to people at home. It's even possible to make one yourself.

However, researchers urge caution. They admit that they still aren't exactly sure how it works, and messing with your brain could have dangerous consequences.

"Prosthetic limbs can now be controlled by the mind"

Exoskeletons and virtual reality

At the 2014 World Cup in Brazil, Miguel Nicolelis from Duke University teamed up with 29-year-old Juliano Pinto to showcase exciting new technology. Pinto is paralysed from the chest down, but with the help of Nicolelis' mind-controlled exoskeleton and a cap to pick up his brainwaves, he was able to stand and kick the official ball.

The next step in Nicolelis' research has been focused on retraining the brain to move the legs – and this time he's using VR. After months of controlling the walking of a virtual avatar with their minds, eight people with spinal-cord injuries have actually regained some movement and feeling in their own limbs.

Electrodes can pick up neural impulses, so paralysed patients are able to control virtual characters with their brain activity

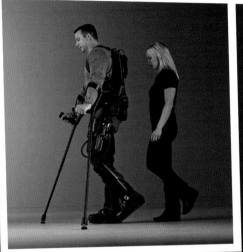

Exosuits can amplify your natural movement, while some models can even be controlled by your mind

Community biology labs

We spoke to Tom Hodder, technical director at London Biological Laboratories Ltd to learn more about public labs and the biohacking movement

Interview bio:
Tom Hodder studied medicinal chemistry and is a biohacker working on open hardware at London Biohackspace.

What is the London Biohackspace?
The London Biohackspace is a biolab at the London Hackspace on Hackney Road. The lab is run by its members, who pay a small monthly fee. In return they can use the facilities for their own experiments and can take advantage of the shared equipment and resources. In general the experiments are some type of microbiology, molecular or synthetic biology, as well as building and repairing biotech hardware.

Who can get involved? Is the lab open to anyone?
Anyone can join up. Use of the lab is subject to a safety induction. There is a weekly meet-up on Wednesdays at 7.30pm, which is open to the public.

Why do you think there is such an interest in biohacking?
Generally, I think that many important problems, such as food, human health, sustainable resources (e.g. biofuels) can be potentially mitigated by greater understanding of the underlying

processes at the molecular biological level. I think that the biohacking community is orientated towards the sharing of these skills and knowledge in an accessible way. Academic research is published, but research papers are not the easiest reading, and the details of commercial research are generally not shared unless it's patented. More recently, much of the technology required to perform these experiments is becoming cheaper and more accessible, so it is becoming practical for biohacking groups to do more interesting experiments.

Where do you see biohacking going in the future?
I think in the short term, the biohacking groups are not yet at an equivalent level to technology and resources to the universities and commercial research institutions. However in the next five years, I expect more open biolabs and biomakerspaces to be set up and the level of sophistication to increase. I think that biohacking groups will continue to perform the service of communicating the potential of synthetic and molecular biology to the general public, and hopefully do that in an interesting way.

Community labs are popping up all over the world, providing amateur scientists with access to biotech equipment

© Thinkstock; Alamy; Ekso Bionics

77

BUILDING FUTURE YOU

A closer look at some of the emerging tech that will allow you to customise your body

Self-improvement is part of human nature, and technology is bringing unprecedented possibilities into reach. Much of the development up until this point has had a medical purpose in mind, including prosthetic limbs for amputees, exoskeletons for paralysis, organs for transplant, and light sensors for the blind. However, with the advent of wearable technology, and a growing community of amateur and professional biotechnology tinkerers, there is increased interest in augmenting the healthy human body.

The first cyborgs already walk among us, fitted with magnetic senses, implanted with microchips, and talking to technology using their nervous systems. At the moment, many devices are experimental, sometimes even homemade and unlicensed. However, the field is opening up, and the possibilities are endless.

So, what does the future hold for a customisable you? Medical implants could monitor, strengthen, heal or replace our organs. We could add extra senses, or improve the ones we already have. And, one day, we might be able to tap straight into the internet with our minds.

Custom-build your body

Technology of the future will offer the opportunity to tinker with the human body like never before

Mind-controlled prosthetics
Using a film of electrode sensors implanted on to the brain, wearers will control bionic limbs just by thinking.

RFID implants
Radio frequency identification chips implanted under the skin store information, open doors and communicate with other technology.

Eye cameras
Retinal implants link light-sensing electronics up to the back of the eye, detecting images and sending the information to the brain.

Smart lenses
Contact lenses fitted with micro-electronics monitor vital medical information, and display an augmented reality overlay on your vision.

Fingertip magnets
Tiny neodymium magnets implanted beneath the skin allow people to lift small magnetic objects, and sense invisible magnetic fields.

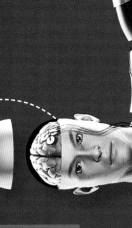

Electronic tattoos

Gold-leaf temporary tattoos can be used as touch sensors, colour-changing indicators, and for Wi-Fi communications.

Smart bandages

Wound dressings will be equipped with sensors to monitor healing and flag up the first signs of infection by turning fluorescent green.

Interchangeable limbs

Advanced prosthetics could give amputees superhuman abilities, and the option to switch between designs to suit the situation.

Bionic organs

Replacement organs will be grown from real human cells in the lab, or reconstructed using synthetic materials and electronics.

Exoskeleton support

Robotic exoskeletons support the wearer's limbs, using hydraulics in place of muscles, and hinges in place of joints.

"Many devices are experimental, sometimes even homemade"

Ekso moves legs in response to upper body movement

The i-limb hand can be moved by gestures, apps, muscle signals or proximity sensors

The Argus implant's camera and transmitter signal to the optic nerve

This RFID chip shows the coiled copper antenna it uses to communicate

Google is developing a contact lens that senses blood sugar by analysing tears

THE BODY AT WORK

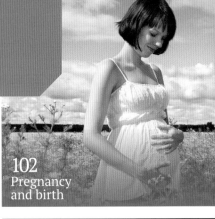

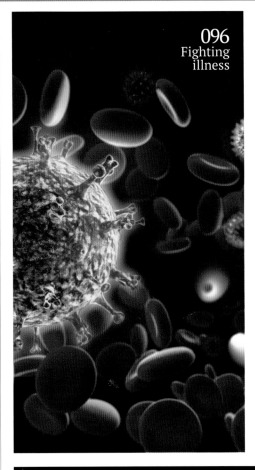

096
Fighting
illness

102
Pregnancy
and birth

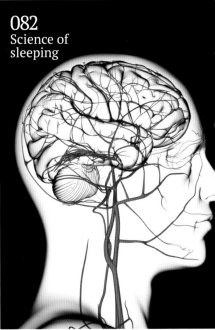

082
Science of
sleeping

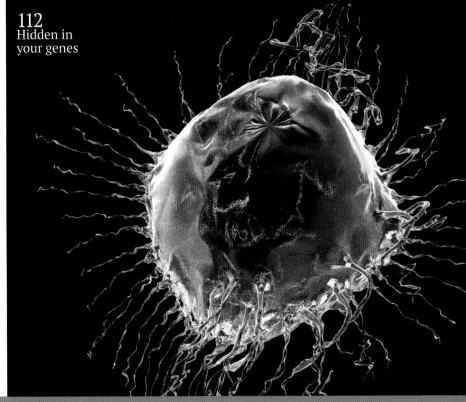

112
Hidden in
your genes

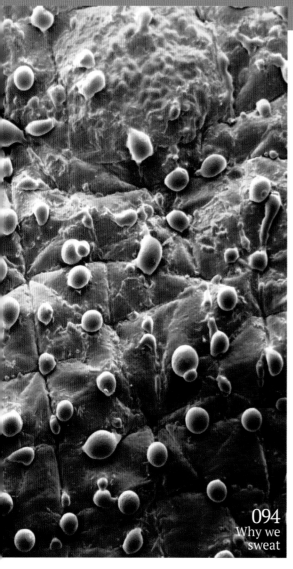

094
Why we
sweat

104
Hyperthermia vs
Hypothermia

106
Bacteria
or virus?

The science of Sleep

Unravelling the mysteries behind insomnia, sleepwalking, dreams and more

We spend around a third of our lives sleeping. It is vital to our survival, but despite years of research, scientists still aren't entirely sure why we do it. The urge to sleep is all-consuming, and if we are deprived of it, we will eventually slip into slumber even if the situation is life-threatening.

Sleep is an essential habit to mammals, birds and reptiles and has been conserved through evolution, despite preventing us from performing tasks such as eating, reproducing and raising young. It is as important as food and, without it, rats will die within two or three weeks – the same period it takes to die of starvation.

There have been many ideas and theories proposed about why humans sleep, from a way to rest after the day's activities or a method for saving energy, to simply a way to fill time until we can be doing something useful. But all of these ideas are somewhat flawed. The body repairs itself just as well when we are sitting quietly, we only save around 100 calories a night by sleeping, and we wouldn't need to catch up on sleep during the day if it were just to fill empty time at night.

One of the major problems with sleep deprivation is a resulting decline in cognitive ability – our brains just don't work properly without sleep. We will find ourselves struggling with memory, learning, planning and reasoning. A lack of sleep can actually have severe effects on our mood and performance of everyday tasks, ranging from irritability, through to long term problems such as an

Theories of why we sleep

Theory 1

Energy conservation

We save around 100 calories per night by sleeping; metabolic rate drops, the digestive system is less active, heart and breathing rates slow, and body temperature drops. However, the calorie-saving equates to just one cup of milk, which from an evolutionary perspective does not seem worth the accompanying vulnerability.

Theory 3

Restoration

One of the major problems with sleep deprivation is a decline in cognitive function, accompanied by a drop in mood, and there is mounting evidence that sleep is involved in restoring the brain. However, there is little evidence to suggest that the body undergoes more repair during sleep compared to rest or relaxation.

Theory 2

Evolutionary protection

An early idea about the purpose of sleep is that it is a protective adaptation to fill time. For example, prey animals with night vision might sleep during the day to avoid being spotted by predators. However, this theory cannot explain why sleep-deprived people fall asleep in the middle of the day.

Theory 4

Memory consolidation

One of the strongest theories regarding sleep is that it helps with consolidation of memory. The brain is bombarded with more information during the day than it is possible to remember, so sleep is used to sort through this information and selectively practise parts that need to be stored.

increased risk of heart disease and even a higher incidence of road traffic accidents.

Sleep can be divided into two broad stages: non-rapid eye movement (NREM), and rapid eye movement (REM) sleep. The vast majority of our sleep, actually around 75 to 80 per cent of it, is NREM, which is characterised by various electrical patterns in the brain known as 'sleep spindles' and high, slow delta waves. When this is occuring, this is the time when we sleep the deepest. Without NREM sleep, our ability to form declarative memories, such as learning to associate pairs of words, can be seriously impaired. Deep sleep is important for transferring short-term memories into long-term storage. Deep sleep is also the time of peak growth hormone release in the body, which is important for cell reproduction and repair.

The purpose of REM sleep is unclear, with the effects of REM sleep deprivation proving less severe than NREM deprivation; for the first two weeks humans report little in the way of ill effects. REM sleep is the period during the night when we have our most vivid dreams, but people dream during both NREM and REM sleep. One curiosity is that during NREM sleep, dreams tend to be more concept-based, whereas REM sleep dreams are a lot more vivid and emotional.

Some scientists argue that REM sleep allows our brains a safe place to practice dealing with situations or emotions that we might not encounter during our daily lives. During REM sleep our muscles are temporarily paralysed, preventing us acting out these emotions. Others think that it might be a way to unlearn memories, or to process unwanted feelings or emotions. Each of these ideas has its flaws, and no one knows the real answer.

We will delve into the science of sleep and attempt to make sense of the mysteries of the sleeping brain.

The sleep cycle

In the night, you cycle through five separate stages of sleep every 90 to 110 minutes

The five stages of sleep can be distinguished by changes in the electrical activity in your brain, measured by electroencephalogram (EEG). The first stage begins with drowsiness as you drift in and out of consciousness, and is followed by light sleep and then by two stages of deep sleep. Your brain activity starts to slow down, your breathing, heart rate and temperature drop, and you become progressively more difficult to wake up. Finally, your brain perks up again, resuming activity that looks much more like wakefulness, and you enter rapid eye movement (REM) sleep – the time when your most vivid dreams occur. This cycle happens several times throughout the night, and each time, the period of REM sleep grows longer.

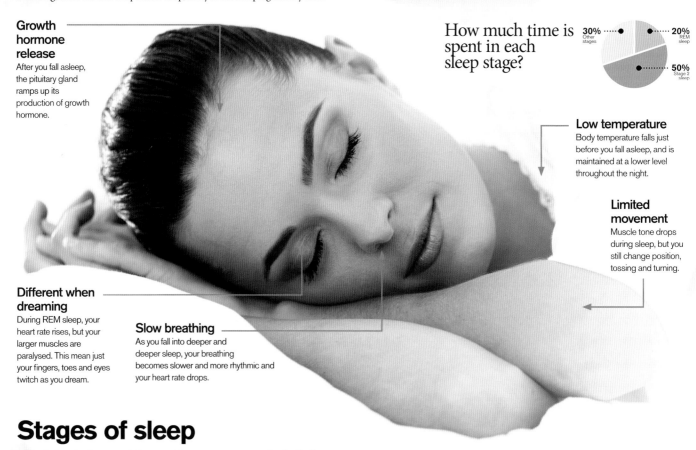

Growth hormone release
After you fall asleep, the pituitary gland ramps up its production of growth hormone.

How much time is spent in each sleep stage?

30% Other stages

20% REM sleep

50% Stage 2 sleep

Low temperature
Body temperature falls just before you fall asleep, and is maintained at a lower level throughout the night.

Limited movement
Muscle tone drops during sleep, but you still change position, tossing and turning.

Different when dreaming
During REM sleep, your heart rate rises, but your larger muscles are paralysed. This mean just your fingers, toes and eyes twitch as you dream.

Slow breathing
As you fall into deeper and deeper sleep, your breathing becomes slower and more rhythmic and your heart rate drops.

Stages of sleep

Not all sleep is the same. There are five separate stages, divided by brain activity

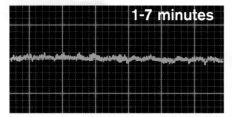

1-7 minutes

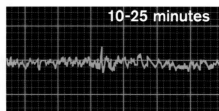

10-25 minutes

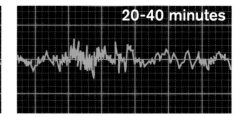

20-40 minutes

1 Drowsiness

During the first stage of sleep you are just drifting off; your eyelids are heavy and your head starts to drop. During this drowsy period, you are easily woken and your brain is still quite active. The electrical activity on an electroencephalogram (EEG) monitor starts to slow down, and the cortical waves become taller and spikier. As the sleep cycle repeats during the night, you re-enter this drowsy half-awake, half-asleep stage.

2 Light sleep

After a few minutes, your brain activity slows further, and you descend into light sleep. On the EEG monitor, this stage is characterised by further slowing in the waves, with an increase in their size and short one- or two-second bursts of activity known as 'sleep spindles'. By the time you are in the second phase of sleep, your eyes stop moving, but you can still be woken up quite easily.

3 Moderate sleep

As you start to enter this third stage, your sleep spindles stop, this in turn is showing that your brain has entered moderate sleep. This is then followed by deep sleep. The trace on the EEG slows still further as your brain produces delta waves with occasional spikes of smaller, faster waves in between. As you progress through stage-three sleep, you become much more difficult to wake up.

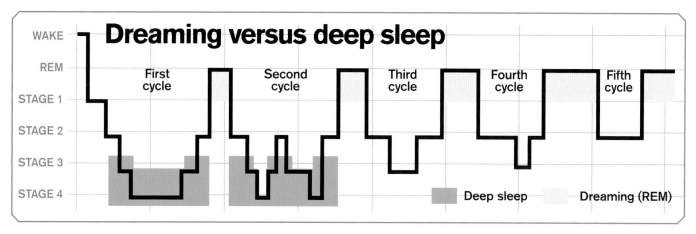

Dreaming versus deep sleep

First cycle · Second cycle · Third cycle · Fourth cycle · Fifth cycle

WAKE · REM · STAGE 1 · STAGE 2 · STAGE 3 · STAGE 4

Deep sleep   Dreaming (REM)

Clearing the mind

The brain is a power-hungry organ; it makes up only two per cent of the total mass of the body, but it uses an enormous 25 per cent of the total energy supply. The question is, how does it get rid of waste? The Nedergaard Lab at the University of Rochester in New York thinks sleep might be a time to clean the brain. The rest of the body relies on the lymphatic drainage system to help remove waste products, but the brain is a protected area, and these vessels do not extend upward into the head. Instead, your central nervous system is bathed in a clear liquid called cerebrospinal fluid (CSF), into which waste can be dissolved for removal. During the day, it remains on the outside, but the lab's research has shown that, during sleep, gaps open up between brain cells and the fluid rushes in, following paths along the outside of blood vessels, sweeping through every corner of the brain and helping to clear out toxic molecules.

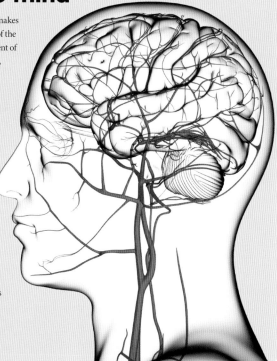

Brain activity

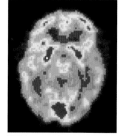

Wide awake
The red areas in this scan show areas of activity in the waking human brain, while the blue areas represent areas of inactivity.

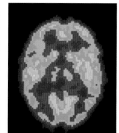

Deep sleep
During the later stages of NREM sleep, the brain is less active, shown here by the cool blue and purple colours that dominate the scan.

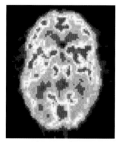

REM (dream) sleep
When we are dreaming, the human brain shows a lot of activity, displaying similar red patterns of activity to the waking brain.

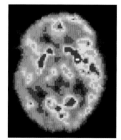

Light sleep
In the first stages of NREM sleep, the brain is less active than when awake, but you remain alert and easy to wake up.

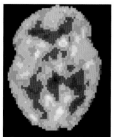

Sleep deprivation
The sleep-deprived brain looks similar to the brain during NREM sleep, showing patterns of inactivity.

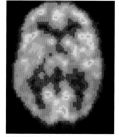

NREM sleep
As you descend through the four stages of NREM sleep, your brain in turn becomes progressively less active.

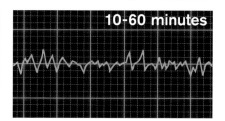

20-40 minutes   10-60 minutes

4 Deep sleep

There is some debate as to whether sleep stages three and four are really separate, or whether they are part of the same phase of sleep. Stage four is the deepest stage of all, and during this time you are extremely hard to wake. The EEG shows tall, slow waves which are known as delta waves; your muscles will relax and your breathing becomes slow and rhythmic, which can lead to snoring.

5 REM sleep

After deep sleep, your brain starts to perk up and its electrical activity starts to resemble the waking brain. This is the period of the night when most dreams happen. Your muscles are temporarily paralysed, and your eyes dart around, giving it the name rapid eye movement (REM) sleep. You cycle through the stages of sleep about every 90 minutes, experiencing between three and five dream periods each night.

Sleep disorders

There are over 100 different disorders that prevent a good night's sleep

Sleep is necessary for our health, so disruptions to the quality or quantity of our sleep can have a serious negative impact on daily life, affecting both physical health and mental wellbeing.

Sleep disorders fall into four main categories: difficulty falling asleep, difficulty staying awake, trouble sticking to a regular sleep pattern and abnormal sleep behaviours. Struggling with falling asleep or staying asleep is known as insomnia, and is one of the most familiar sleep disorders; around a third of the population will experience it during their lifetime. Difficulty staying awake, or hypersomnia, is less common. The best-known example is narcolepsy, which is when sufferers experience excessive daytime sleepiness, accompanied by uncontrollable short periods of sleep during the day. Trouble sticking to a regular sleeping pattern can either be caused by external disruption to normal day-to-day rhythms, for example by jet lag or shift work. It can also be the result of an internal problem with the part of the brain responsible for setting the body clock.

Abnormal sleep behaviours include problems like night terrors, sleepwalking and REM-sleep behaviour disorder. Night terrors and sleepwalking most commonly affect children, and tend to resolve themselves with age, but other sleep behaviours persist into adulthood. In REM-sleep behaviour disorder, the normal muscle paralysis that accompanies dreaming fails, and people begin to act out their dreams.

Treatment for different sleep disorders varies depending on the particular problem, and sometimes it can even be as simple as making the individuals bedroom environment more conducive to restful sleep.

"Treatment for different sleep disorders varies"

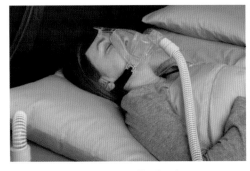

A continuous positive airway pressure (CPAP) machine pumps air into a close-fitting mask, preventing the airway from collapsing

Sleepwalking

Sleepwalkers can perform complicated actions while in deep NREM sleep

Sleepwalking affects between one and 15 per cent of the population, and is much more common in children than in adults, tending to happen less and less after the age of 11 or 12. Sleepwalkers might just sit up in their bed, but can sometimes perform complex behaviours, such as walking, getting dressed, cooking, or even driving a car. Although sleepwalkers seem to be acting out their dreams, sleepwalking tends to occur during the deep-sleep phase of NREM sleep and not during REM sleep.

Sleep apnoea

Sleep apnoea is a dangerous sleep disorder. It is when the walls of the airways relax so much during the night that breathing is interrupted for ten seconds or more, restricting the supply of oxygen to the brain. The lack of oxygen initiates a protective response, pulling the sufferer out of deep sleep to protect them from damage. This can cause people to wake up, but often it will just put them into a different sleep stage, interrupting their rest and causing feelings of tiredness the next day.

Loud breathing
People suffering with sleep apnoea often snore, gasp and breathe loudly as they struggle for air during the night.

Waking up
The low oxygen level in the blood triggers the brain to wake up in an attempt to fix the obstruction.

Lack of oxygen
If the airway is obstructed for ten seconds or more, the amount of oxygen reaching the brain drops.

Muscle collapse
The muscles supporting the tongue, tonsils and soft palate relax during sleep, causing the throat to narrow.

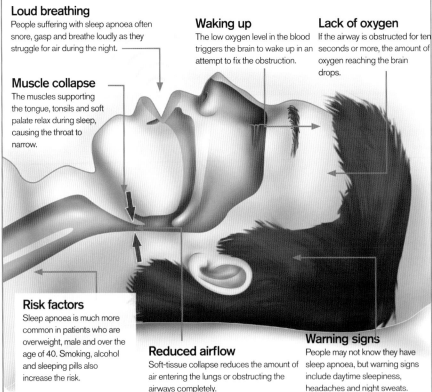

Risk factors
Sleep apnoea is much more common in patients who are overweight, male and over the age of 40. Smoking, alcohol and sleeping pills also increase the risk.

Reduced airflow
Soft-tissue collapse reduces the amount of air entering the lungs or obstructing the airways completely.

Warning signs
People may not know they have sleep apnoea, but warning signs include daytime sleepiness, headaches and night sweats.

Narcolepsy

Narcolepsy is a chronic condition that causes people to suddenly fall asleep during the daytime. In the United States, it affects one in every 3,000 people. Narcoleptics report excessive amounts of daytime sleepiness, accompanied by a lack of energy and impaired ability to concentrate. They fall asleep involuntarily for periods lasting just a few seconds at a time, and some can continue to perform tasks such as writing, walking, or even driving during these microsleeps. In 70 per cent of cases, narcolepsy is also accompanied by cataplexy, where the muscles go limp and become difficult to control. It has been linked to low levels of the neurotransmitter hypocretin, which is responsible for promoting wakefulness in the brain.

People with narcolepsy fall asleep involuntarily during the day

Insomnia

Insomniacs have difficulty falling asleep or staying asleep. Sufferers can wake up during the night, wake up unusually early in the morning, and report feeling tired and drained during the day. Stress is thought to be one of the major causes of this sleep disruption, but it is also associated with mental health problems like depression, anxiety and psychosis, and also with underlying medical conditions that range from lung problems to hormone imbalances. After underlying causes have been ruled out, management of insomnia generally involves improving 'sleep hygiene' by sticking to regular sleep patterns, avoiding caffeine in the evening and keeping the bedroom free from light and noise at night.

One in three people in the UK will experience insomnia in their lifetime

Sleep studies

The most common type of sleep study is a polysomnogram (PSG), which is an overnight test performed in a specialist sleep facility. Electrodes are placed on the chin, scalp and eyelids to monitor brain activity and eye movement, while pads are placed on the chest to track heart rate and breathing. Their blood pressure is also monitored throughout the night, and the amount of oxygen in the bloodstream can be tracked using a device worn on the finger. The equipment monitors how long it takes a patient to fall asleep, and then to follow their brains and bodies as they move through each of the five different sleep stages.

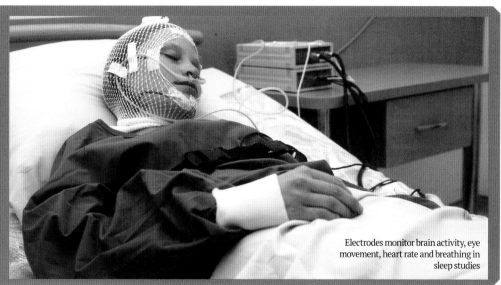

Electrodes monitor brain activity, eye movement, heart rate and breathing in sleep studies

How to get a good night's sleep

Understanding your biological clock is the key to a healthy night's sleep

Your body is driven by an internal circadian master clock known as the suprachiasmatic nucleus, which is set on a time scale of roughly 24 hours. This biological clock is set by sunlight; blue light hits special receptors in your eyes, which feed back to the master clock and on to the pineal gland. This suppresses the production of the sleep hormone melatonin and tells your brain that it is time to wake up.

Disruptions in light exposure can play havoc with your sleep, so it is important to ensure that your bedroom is as dark as possible. Many electronic devices produce enough light to reset your biological clock, and using backlit screens late at night can confuse your brain, preventing the production of melatonin and delaying your sleep.

Ensuring you see sunlight in the morning can help to keep your circadian clock in line, and sticking to a regular sleep schedule, even at the weekends, helps to keep this rhythm regular.

Another important factor in a good night's sleep is the process of winding down before bed. Certain stimulants such as caffeine and nicotine will actually keep your brain alert and can seriously disrupt your attempts to sleep. Even depressants like alcohol can have a negative effect; even though it calms the brain, it interferes with normal sleep cycles, preventing proper, restorative, deep and REM sleep.

The blue light from televisions, mobile phones and computer screens disrupts your circadian rhythm

The dangers of sleep deprivation

Lack of sleep doesn't just make you tired – it can have dangerous unseen effects

1 IMPAIRED JUDGEMENT

Sleep deprivation impacts your visual working memory, making it hard to distinguish between relevant and irrelevant stimuli, affecting emotional intelligence, behaviour and stress management.

2 WEIGHT GAIN

Sleep deprivation affects the levels of hormones involved in regulating appetite. Levels of leptin (the hormone that tells you how much stored fat you have) drop, and levels of the hunger hormone ghrelin rise.

3 RAISED BLOOD PRESSURE

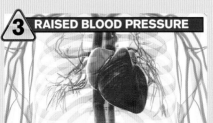

Poor sleep can raise blood pressure, and in the long term is associated with an increased risk of diseases such as coronary heart disease and stroke. This danger is increased in people with sleep apnoea.

4 INCREASED ACCIDENTS

In the USA it is estimated that 100,000 road accidents each year are the result of driver fatigue, and over a third of drivers have even admitted to falling asleep behind the wheel.

5 MOOD DISORDERS

Mental health problems are linked to sleep disorders, and having sleep deprivation can play havoc with neurotransmitters in the brain, mimicking the symptoms of depression, anxiety and mania.

6 HALLUCINATIONS

Severe sleep deprivation can lead to hallucinations – seeing things that aren't really there. In rare cases, it can lead to temporary psychosis or symptoms that resemble paranoid schizophrenia.

Sleep myths debunked

The science behind five of the most common myths relating to sleep

"Counting sheep helps you sleep"

MYTH DEBUNKED

This myth was put to the test by the University of Oxford, who challenged insomniacs to either count sheep, imagine a relaxing scene, or do nothing as they tried to fall asleep. When they imagined a relaxing scene, the participants fell asleep an average of 20 minutes earlier than when they tried either of the other two methods.

"Yawning wakes you up"

MYTH DEBUNKED

Yawning has long been associated with tiredness and was fabled to provide more oxygen to a sleepy brain, but this is not the case. New research suggests that we actually yawn to cool our brains down, using a deep intake of breath to keep the brain running at its optimal temperature.

"Teenagers are lazy"

MYTH DEBUNKED

Sleep habits start to change just before puberty, and between the ages of ten and 25, people need around nine hours of sleep every night. Teens can also experience a shift in their circadian rhythm, called sleep phase delay, pushing back their natural bedtime by around two hours, and encouraging them to sleep in.

"You should never wake a sleepwalker"

Many people have heard that waking a sleepwalker might kill them, but there is little truth behind these tales. Waking a sleepwalker can leave them confused and disorientated, but the act of sleepwalking in itself can be much more dangerous. Gently guiding a sleepwalker back to their bed is the safest option, but waking them carefully shouldn't do any harm.

MYTH DEBUNKED

"Cheese gives you nightmares"

MYTH DEBUNKED

The British Cheese Board conducted a study in an attempt to debunk this myth by feeding 20g (0.7oz) of cheese to 200 volunteers every night for a week and asking them to record their dreams. There were no nightmares, but strangely 75 per cent of men and 85 per cent of the women who ate Stilton reported vivid dreams.

SLEEP STATS

What are the most common sleeping positions?

41% Foetus

15% Log

13% Yearner

8% Soldier

7% Freefaller

5% Starfish

How does sleep time vary with age?

16 hours INFANTS

9 hours TEENS

7 hours ADULTS

What keeps the UK up at night?

67% Discomfort

36% Noise

13% Partner

34% Temperature

19% Light

Which country sleeps the longest?

Canada **7h03**

UK **6h49**

6h22

USA **6h31**

7h06

Germany **7h01**

What do people dream about?

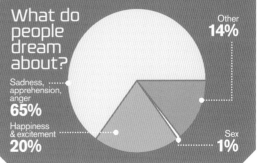

Other **14%**

Sadness, apprehension, anger **65%**

Happiness & excitement **20%**

Sex **1%**

Human digestion

How does food get turned into energy?

The digestive system is a group of organs that process food into energy that the human body can use to operate. It is an immensely complex system that stretches all the way between the mouth and the anus.

Primary organs that make up the system are the mouth, oesophagus, stomach, small intestine, large intestine and the anus. Each organ has a different function so that the maximum amount of energy is gained from the food, and the waste can be safely expelled from the body. Secondary organs, such as the liver, pancreas and gall bladder, aid the digestive process alongside mucosa cells, which line all hollow organs and produce a secretion which helps the food pass smoothly through them. Muscle contractions called peristalsis also help to push the food throughout the system.

The whole digestive process starts when food is taken into the body through the mouth. Mastication (chewing) breaks down the food into smaller pieces and saliva starts to break starch in these pieces of food into simpler sugars as they are swallowed and move into the oesophagus. Once the food has passed through the oesophagus, it passes into the stomach. It can be stored in the stomach for up to four hours.

The stomach will eventually mix the food with the digestive juices that it produces, which will break down the food further into simpler molecules. These molecules then move into the small intestine slowly, where the final stage of chemical breakdown occurs through exposure to juices and enzymes released from the pancreas, liver and glands in the small intestine. All the nutrients are then absorbed through the intestinal walls and transported around the body through the blood stream.

After all nutrients have been absorbed from food through the small intestine, resulting waste material, including fibre and old mucosa cells, is then pushed into the large intestine where it will remain until expelled by a bowel movement.

"Nutrients are then absorbed through the intestinal walls and transported around the body"

Mouth
This is where food enters the body and first gets broken into more manageable pieces. Saliva is produced in the glands and starts to break down starch in the food.

Oesophagus
The oesophagus passes the food into the stomach. At this stage, it has been broken down through mastication and saliva will be breaking down starch.

Large intestine
The colon, as the large intestine is also known, is where waste material will be stored until expelled from the digestive system through the rectum.

Small intestine
Nutrients that have been released from food are absorbed into the blood stream so they can be transported to where they are needed in the body through the small intestine wall. Further breaking down occurs here with enzymes from the liver and pancreas.

How your body digests food

Many different organs are involved in the digestion process

Rectum
This is where waste material (faeces) exits the digestive system.

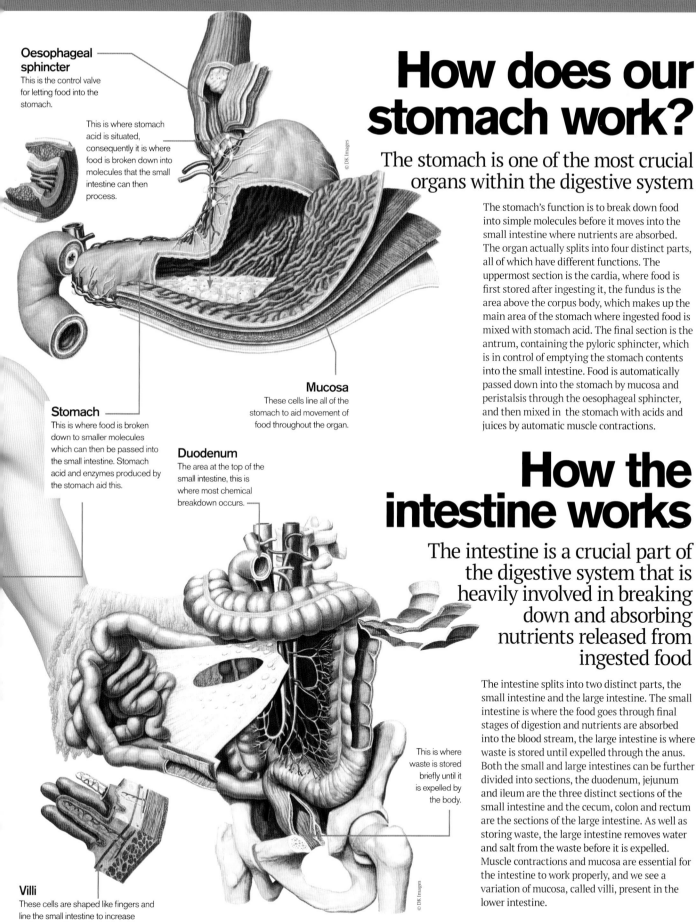

Oesophageal sphincter
This is the control valve for letting food into the stomach.

This is where stomach acid is situated, consequently it is where food is broken down into molecules that the small intestine can then process.

Mucosa
These cells line all of the stomach to aid movement of food throughout the organ.

Stomach
This is where food is broken down to smaller molecules which can then be passed into the small intestine. Stomach acid and enzymes produced by the stomach aid this.

Duodenum
The area at the top of the small intestine, this is where most chemical breakdown occurs.

This is where waste is stored briefly until it is expelled by the body.

Villi
These cells are shaped like fingers and line the small intestine to increase surface area for nutrient absorption.

© DK Images

How does our stomach work?

The stomach is one of the most crucial organs within the digestive system

The stomach's function is to break down food into simple molecules before it moves into the small intestine where nutrients are absorbed. The organ actually splits into four distinct parts, all of which have different functions. The uppermost section is the cardia, where food is first stored after ingesting it, the fundus is the area above the corpus body, which makes up the main area of the stomach where ingested food is mixed with stomach acid. The final section is the antrum, containing the pyloric sphincter, which is in control of emptying the stomach contents into the small intestine. Food is automatically passed down into the stomach by mucosa and peristalsis through the oesophageal sphincter, and then mixed in the stomach with acids and juices by automatic muscle contractions.

How the intestine works

The intestine is a crucial part of the digestive system that is heavily involved in breaking down and absorbing nutrients released from ingested food

The intestine splits into two distinct parts, the small intestine and the large intestine. The small intestine is where the food goes through final stages of digestion and nutrients are absorbed into the blood stream, the large intestine is where waste is stored until expelled through the anus. Both the small and large intestines can be further divided into sections, the duodenum, jejunum and ileum are the three distinct sections of the small intestine and the cecum, colon and rectum are the sections of the large intestine. As well as storing waste, the large intestine removes water and salt from the waste before it is expelled. Muscle contractions and mucosa are essential for the intestine to work properly, and we see a variation of mucosa, called villi, present in the lower intestine.

© DK Images

Human respiration

Respiration is crucial to an organism's survival. The process of respiration is the transportation of oxygen from the air that surrounds us into the tissue cells of our body so that energy can be broken down

The primary organs used for respiration in humans are the lungs. Humans have two lungs, with the left lung being divided into two lobes and the right into three. Lungs have between 300–500 million alveoli, which is actually where gas exchange occurs.

Respiration of oxygen breaks into four main stages: ventilation, pulmonary gas exchange, gas transportation and peripheral gas exchange. Each stage is crucial in getting oxygen to the body's tissue, and removing carbon dioxide. Ventilation and gas transportation need energy to occur, as the diaphragm and the heart are used to facilitate these actions, whereas gas exchanging is passive. As air is drawn into the lungs at a rate of between 10-20 breaths per minute while resting, through either your mouth or nose by diaphragm contraction, and travels through the pharynx, then the larynx, down the trachea, and into one of the two main bronchial tubes. Mucus and cilia keep the lungs clean by catching dirt particles and sweeping them up the trachea.

When air reaches the lungs, oxygen is diffused into the bloodstream through the alveoli and carbon dioxide is diffused from the blood into the lungs to be exhaled. Diffusion of gases occurs because of differing pressures in the lungs and blood. This is also the same when oxygen diffuses into tissue around the body. When blood has been oxygenated by the lungs, it is transferred around the body to where it is most needed in the bloodstream. If the body is exercising, the breathing rate increases and, consequently, so does the heart rate to ensure that oxygen reaches tissues that need it. Oxygen is then used to break down glucose to provide energy for the body. This happens in the mitochondria of cells. Carbon dioxide is one of the waste products of this, which is why we get a build up of this gas in our body that needs to be transported back into the lungs to then be exhaled.

The body can also respire anaerobically, but this produces far less energy and instead of producing co2 as a byproduct, lactic acid is produced. The body then takes time to break this down after exertion has finished as the body has a so-called oxygen debt.

2. Pharynx
This is part of both the respiratory and digestive system. A flap of connective tissue called the epiglottis closes over the trachea to stop choking when an individual takes food into their body.

1. Nasal passage/ oral cavity
These areas are where air enters into the body so that oxygen can be transported into and around the body to where it's needed. Carbon dioxide also exits through these areas.

5. Alveoli
The alveoli are tiny little sacs which are situated at the end of tubes inside the lungs and are in direct contact with blood. Oxygen and carbon dioxide transfer to and from the blood stream through the alveoli.

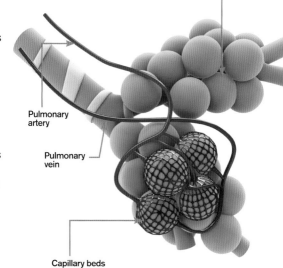

Pulmonary artery

Pulmonary vein

Capillary beds

How our lungs work
Lungs are the major respiratory organ in humans

How do we breathe?

The intake of oxygen into the body is complex

Breathing is not something that we have to think about, and indeed is controlled by muscle contractions in our body. Breathing is controlled by the diaphragm, which contracts and expands on a regular, constant basis. When it contracts, the diaphragm pulls air into the lungs by a vacuum-like effect. The lungs expand to fill the enlarged chest cavity and air is pulled right through the maze of tubes that make up the lungs to the alveoli at the ends, which are the final branching. The chest will be seen to rise because of this lung expansion. Alveoli are surrounded by various blood vessels, and oxygen and carbon dioxide are then interchanged at this point between the lungs and the blood. Carbon dioxide removed from the blood stream and air that was breathed in but not used is then expelled from the lungs by diaphragm expansion. Lungs deflate back to a reduced size when breathing out.

3. Trachea
Air is pulled into the body through the nasal passages and then passes into the trachea.

4. Bronchial tubes
These tubes lead to either the left or the right lung. Air passes through these tubes into the lungs, where they pass through progressively smaller and smaller tubes until they reach the alveoli.

6. Ribs
These provide protection for the lungs and other internal organs situated in the chest cavity.

© DK Images

Chest cavity
This is the space that is protected by the ribs, where the lungs and heart are situated. The space changes as the diaphragm moves.

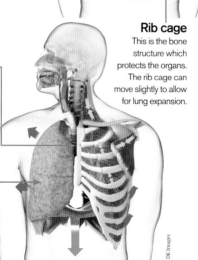

© DK Images

Lungs
Deoxygenated blood arrives back at the lungs, where another gas exchange occurs at the alveoli. Carbon dioxide is removed and oxygen is placed back into the blood.

Diaphragm
This is a sheet of muscle situated at the bottom of the rib cage which contracts and expands to draw air into the lungs.

Heart
The heart pumps oxygenated blood away from the lungs, around the body to tissue, where oxygen is needed to break down glucose into a usable form of energy.

Tissue
Oxygen arrives where energy is needed, and a gas exchange of oxygen and carbon dioxide occurs so that aerobic respiration can occur within cells.

Rib cage
This is the bone structure which protects the organs. The rib cage can move slightly to allow for lung expansion.

© DK Images

Why do we need oxygen?

We need oxygen to live as it is crucial for the release of energy within the body

Although we can release our energy through anaerobic respiration temporarily, this method is inefficient and creates an oxygen debt that the body must repay after excess exercise or exertion has ceased. If oxygen supply is cut off for more than a few minutes, an individual will die. Oxygen is pumped around the body to be used in cells that need to break down glucose so that energy is provided for the tissue. The equation that illustrates this is:

$$C_6H_{12}O_6 + 6O_2 = 6CO_2 + 6H_2O + energy$$

During heavy cardiovascular exercise, extra oxygen is required

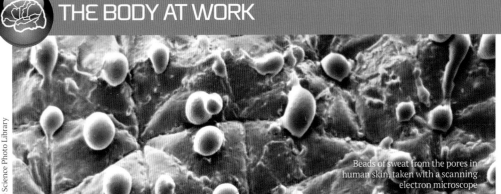

© Science Photo Library

Beads of sweat from the pores in human skin, taken with a scanning electron microscope

Why do we sweat?

As your doctor may tell you, it's glandular…

Sweat is produced by dedicated sweat glands, and is a mechanism used primarily by the body to reduce its internal temperature. There are two types of sweat gland in the human body, the eccrine gland and the apocrine gland. The former regulates body temperature, and is the primary source of excreted sweat, with the latter only secreting under emotional stresses, rather than those involved with body dehydration.

Eccrine sweat glands are controlled by the sympathetic nervous system and, when the internal temperature of the body rises, they secrete a salty, water-based substance to the skin's surface. This liquid then cools the skin and the body through evaporation, storing and then transferring excess heat into the atmosphere.

Both the eccrine and apocrine sweat glands only appear in mammals and, if active over the majority of the animal's body, act as the primary thermoregulatory device. Certain mammals such as dogs, cats and sheep only have eccrine glands in specific areas – such as paws and lips – warranting the need for them to pant in order to control their temperature.

Pore
Sweat is released directly into the dermis via the secretary duct, which then filters through the skin's pores to the surface.

Skin
Once the sweat is on the skin's surface, its absorbed moisture evaporates, transferring the heat into the atmosphere.

Secretary duct
Secreted sweat travels up to the skin via this duct.

Nerve fibres
Deliver messages to glands to produce sweat when the body temperature rises.

Secretary part
This is where the majority of the gland's secretary cells can be located.

Dehydration

What happens if we don't drink enough?

Just by breathing, sweating and urinating, the average person loses ten cups of water a day. With H2O making up as much as 75 per cent of our body, dehydration is a frequent risk. Water is integral in maintaining our systems and it performs limitless functions.

Essentially, dehydration strikes when your body takes in less fluid than it loses. The mineral balance in your body becomes upset with salt and sugar levels going haywire. Enzymatic activity is slowed, toxins accumulate more easily and your breathing can even become more difficult as the lungs are having to work harder.

Babies and the elderly are most susceptible as their bodies are not as resilient as others. It has been recommended to have eight glasses of water or two litres a day. More recent research is undecided as to how much is exactly needed.

Too much H_2O?

Hydration is all about finding the perfect balance. Too much hydration is just as harmful as well as drinking too little; this is known as water intoxication. If an individual has too much liquid in their body, nutrients such as electrolytes and sodium are diluted and the body suffers. Your cells will begin to bloat and expand to such a point that they can even burst, and it can be fatal if untreated with IV fluids containing electrolytes.

Dangers of dehydration

How does a lack of water vary from mild to fatal?

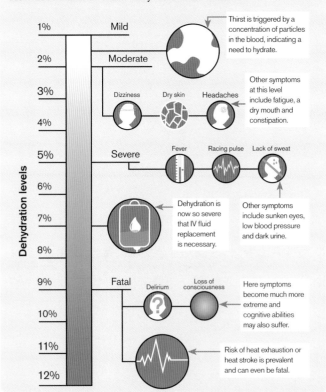

Dehydration levels

1% — Mild

Thirst is triggered by a concentration of particles in the blood, indicating a need to hydrate.

2% — Moderate

Other symptoms at this level include fatigue, a dry mouth and constipation.

3% / 4% — Dizziness / Dry skin / Headaches

5% — Severe — Fever / Racing pulse / Lack of sweat

Other symptoms include sunken eyes, low blood pressure and dark urine.

6%

7% — Dehydration is now so severe that IV fluid replacement is necessary.

8%

9% — Fatal — Delirium / Loss of consciousness

Here symptoms become much more extreme and cognitive abilities may also suffer.

10%

11%

Risk of heat exhaustion or heat stroke is prevalent and can even be fatal.

12%

How wounds heal

It takes an army of cells to repair cuts and scrapes

Wound healing happens in four key stages: haemostasis, inflammation, proliferation and remodelling. Haemostasis means 'blood halting' in Greek, and is the first crucial part in closing a wound. The body's first line of defence is to constrict the blood vessels in the affected area to minimise blood loss. Platelets then start to stick to the exposed tissue, becoming activated and encouraging more and more platelets to clump together to plug the gap.

Once this plug is in place, a mesh of fibrin fibres starts to form around it, trapping passing blood cells and forming a sturdy clot that holds the wound closed until it can be repaired. This process only takes a matter of minutes, and once the bleeding stops, the local blood vessels dilate again, allowing immune cells to reach the area and begin the necessary repairs. This stage is called inflammation.

White blood cells clear up dead cells, get rid of damaged tissue and chase down any pathogens that have entered through the wound and destroy them by phagocytosis (ingesting them). They also prepare the area for the repair phase, which is known as proliferation.

With the encouragement of the immune system, long, spindle-shaped cells called fibroblasts start rebuilding the collagen scaffolding that holds healthy tissue together. On top of the wound, epithelial cells begin dividing and migrating to cover the gap. New blood vessels start to form and, as the tissue heals, myofibroblasts tug at the edges of the wound to close the hole. Once this stage is complete, it's time for remodelling. The scaffolding built by the fibroblasts is rearranged, and any unneeded cells that were made during the healing process are safely removed.

Halting infection
The immune system rushes in to prevent pathogens entering through an open wound

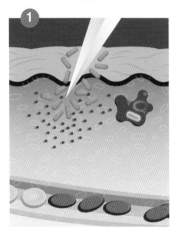 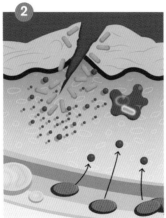 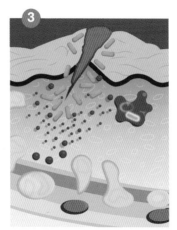 

1. Injury
Blood vessels in the local area instantly constrict, reducing blood loss and limiting the chances of anything entering the bloodstream.

2. Clotting
A plug of platelets starts to form. Clotting factors transform it into a strong network of fibres and trapped blood cells, barricading the breach.

3. Immune response
Immune cells flood into the area, chasing down and killing any bacteria and cleaning away dead cells and damaged tissue.

4. Repair
Immune cells encourage other cells to begin repairs. The support network under the skin is regenerated, and new cells grow over the wound.

5. Macrophage
Macrophage means 'big eater'. These cells are responsible for clearing away pathogens and debris.

6. Immune arrival
Immune cells squeeze out of the blood vessels and into the tissue in a process called extravasation.

7. Granulation
New tissue starts to form at the site of the injury, disordered at first before gradually becoming orderly.

8. Bacteria
An open wound allows pathogens like bacteria to get inside the body.

9. Following the trails
Immune cells are attracted to the site of the wound by a trail of chemical signals.

10. Scarring
If the dermis (deep layer of skin) has been damaged, new collagen fibres form to mend it, creating a scar.

© SPL, Illustration by Ed Crooks

How your immune system works

Physical defences

Human anatomy subscribes to the notion that good fences make good neighbours. Your skin, made up of tightly packed cells and an antibacterial oil coating, keeps most pathogens from ever setting foot in body. Your body's openings are well-fortified too. Pathogens that you inhale face a wall of mucus-covered membranes in your respiratory tract, optimised to trap germs. Pathogens that you digest end up soaking in a bath of potent stomach acid. Tears flush pathogens out of your eyes, dousing bacteria with a harsh enzyme for good measure.

Your body is locked in a constant war against a vicious army

It's true: while you're simply sitting around watching TV, trillions and trillions of foreign invaders are launching a full scale assault on the trillions of cells that constitute 'you'. Collectively known as pathogens, these attackers include bacteria, single-celled creatures that live to eat and reproduce; protists, larger single-cell organisms; viruses, packets of genetic information that take over host cells and replicate inside them; and fungi, a type of plant life.

Bacteria and viruses are by far the very worst offenders. Dangerous bacteria release toxins in the body that cause diseases such as E. coli, anthrax, and the black plague. The cell damage from viruses causes measles, the flu and the common cold, among numerous other diseases.

Just about everything in our environment is teeming with these microscopic intruders, including you. The bacteria in your stomach alone outnumber all the cells in your body, ten-to-one. Yet, your microscopic soldiers usually win against pathogens, through a combination of sturdy barriers, brute force, and superior battlefield intelligence, collectively dubbed the immune system.

The adaptive immune system

Fighting the good fight, and white blood cells are right on the front line...

When a pathogen is tough, wily, or numerous enough to survive various non-specific defences, it's down to the incredibly adaptive immune system to clean up the mess. The key forces in the adaptive immune system are white blood cells which are called lymphocytes. Unlike their macrophage cousins, these lymphocytes are engineered to attack only one specific type of pathogen. There are two types of lymphocytes: B-cells and T-cells.

These cells join the action when macrophages pass along information about the invading pathogen, through chemical messages called interleukins. After engulfing a pathogen, a macrophage communicates details about the pathogen's antigens – telltale molecules that actually characterise particular pathogens. Based on this information, the immune system identifies specific B-cells and T-cells equipped to recognise and battle the pathogen. Once they are successfully identified, these cells rapidly reproduce, assembling an army of cells that are equipped to take down the attacker.

The B-cells flood your body with antibodies, molecules that either disarm a specific pathogen or bind to it, marking it as a target for other white blood cells. When T-cells find their target, they lock on and release toxic chemicals that will destroy it. T-cells are especially adept at destroying your body's cells that are infected with a dangerous virus.

This entire process takes several days to get going and may take even longer to conclude. All the while, the raging battle can make you feel terrible. Fortunately, the immune system is engineered to learn from the past. While your body is producing new B-cells and T-cells to fight the pathogens, it also produces memory cells – copies of the B-cells and T-cells, which stay in the system after the pathogen is defeated. The next time that pathogen shows up in your body, these memory cells help launch a counter-attack much more quickly. Your body can wipe out the invaders before any infection takes hold. In other words, you develop immunity.

Vaccines accomplish exactly the same thing as this by simply giving you just enough pathogen exposure for you to develop memory cells, but not enough to make you sick.

2. Bacterium antigen
These distinctive molecules allow your immune system to recognise that the bacterium is something other than a body cell.

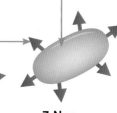

4. Engulfed bacterium
During the initial inflammation reaction, a macrophage engulfs the bacterium.

3. Macrophage
These white blood cells engulf and digest any pathogens they come across.

1. Bacterium
Any bacteria that enter your body have characteristic antigens on their surface.

5. Presented bacterium antigen
After engulfing the bacterium, the macrophage 'presents' the bacterium's distinctive antigens, communicating the presence of the specific pathogen to B-cells.

6. Matching B-cell
The specific B-cell that recognises the antigen, and can help defeat the pathogen, receives the message.

7. Non-matching B-cells
Other B-cells, engineered to attack other pathogens, don't recognise the antigen.

9. Memory cell
The matching B-cell also replicates to produce memory cells, which will rapidly produce copies of itself if the specific bacteria ever returns.

Non-specific defences

As good as your physical defence system is, pathogens do creep past it regularly. Your body initially responds with counterattacks known as non-specific defences, so named because they don't target a specific type of pathogen.

After a breech – bacteria rushing in through a cut, for example – cells release chemicals called inflammatory mediators. This triggers the chief non-specific defence, known as inflammation. Within minutes of a breach, your blood vessels dilate, allowing blood and other fluid to flow into the tissue around the cut.

The rush of fluid in inflammation carries various types of white blood cells, which get to work destroying intruders. The biggest and toughest of the bunch are macrophages, white blood cells with an insatiable appetite for foreign particles. When a macrophage detects a bacterium's telltale chemical trail, it grabs the intruder, engulfs it, takes it apart with chemical enzymes, and spits out the indigestible parts. A single macrophage can swallow up about 100 bacteria before its own digestive chemicals destroy it from within.

How B-cells attack

B-cells target and destroy specific bacteria and invaders

11. Phagocyte
White blood cells called phagocytes recognise the antibody marker, engulf the bacteria, and digest them.

10. Antibodies
The plasma cells release antibodies, which disable the bacteria by latching on to their antigens. The antibodies also mark the bacteria for destruction.

8. Plasma cell
The matching B-cell replicates itself, creating many plasma cells to fight all the bacteria of this type in the body.

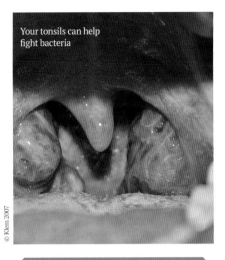

Your tonsils can help fight bacteria

Disorders of the immune system

Who watches the watchmen?

The immune system is a powerful set of defences, so when it malfunctions, it can do as much harm as a disease. Allergies are the result of an overzealous immune system. In response to something that is relatively benign, like pollen for example, the immune system will trigger excessive measures to expel the pathogen. In extreme cases, allergies cause anaphylactic shock, which is a potentially deadly drop in blood pressure, sometimes accompanied by breathing difficulty and loss of consciousness. In autoimmune disorders such as rheumatoid arthritis, the immune system fails to recognise the body's own cells and attacks them.

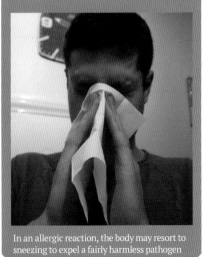

In an allergic reaction, the body may resort to sneezing to expel a fairly harmless pathogen

1. Tonsils
Lymphoid tissue loaded with lymphocytes, which attack bacteria that get into the body through your nose or mouth.

2. Left subclavian vein
One of two large veins that serve as the re-entry point for lymph returning to the bloodstream.

3. Right lymphatic duct
Passageway leading from lymph vessels to the right subclavian vein.

4. Right subclavian vein
The second of the two subclavian veins, this one taking the opposite path to its twin.

5. Spleen
An organ that houses white blood cells that attack pathogens in the body's bloodstream.

10. Lymph vessels
Lymph collects in tiny capillaries, which expand into larger vessels. Skeletal muscles move lymph through these vessels, back into the bloodstream.

6. Lymph node cluster
Located along lymph vessels throughout the body, lymph nodes filter lymph as it makes its way back into the bloodstream.

7. Left lymphatic duct
Passageway leading from lymph vessels to the left subclavian vein.

8. Thymus gland
Organ that provides area for lymphocytes produced by bone marrow to mature into specialised T-cells.

9. Thoracic duct
The largest lymph vessel in the body.

11. Peyer's patch
Nodules of lymphoid tissue supporting white blood cells that battle pathogens in the intestinal tract.

12. Bone marrow
The site of all white blood cell production.

The lymphatic system

The lymphatic system is a network of organs and vessels that collects lymph – fluid that has drained from the bloodstream into bodily tissues – and returns it to your bloodstream. It also plays a key role in your immune system, filtering pathogens from lymph and providing a home-base for disease-fighting lymphocytes.

Lymph nodes explained

Lymph nodes filter out pathogens through your lymph vessels

Your immune system depends on these .04-1-inch swellings to fight all manner of pathogens. As lymph makes its way through a network of fibres in the node, white blood cells filter it, destroying any pathogens they find.

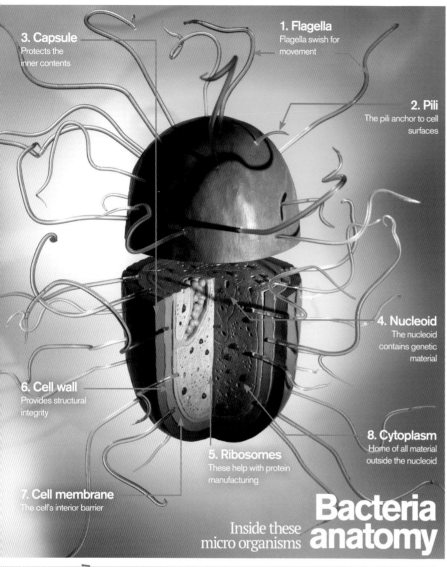

3. Capsule
Protects the inner contents

1. Flagella
Flagella swish for movement

2. Pili
The pili anchor to cell surfaces

4. Nucleoid
The nucleoid contains genetic material

6. Cell wall
Provides structural integrity

8. Cytoplasm
Home of all material outside the nucleoid

5. Ribosomes
These help with protein manufacturing

7. Cell membrane
The cell's interior barrier

Inside these micro organisms

Bacteria anatomy

Know your enemy:
Bacteria

Bacteria are the smallest and, by far, the most populous form of life on Earth. Right now, there are trillions of the single-celled creatures crawling on and in you. In fact, they constitute about four pounds of your total body weight. To the left is a look at bacteria anatomy…

What is HIV…
… and how does it affect the immune system?

The human immunodeficiency virus (HIV) is a retrovirus (a virus carrying ribonucleic acid, or RNA as it's known), transmitted through bodily fluids. Like other deadly viruses, HIV invades cells and multiplies rapidly inside. Specifically, HIV infects cells with CD4 molecules on their surface, which includes infection-fighting helper T-cells. HIV destroys the host cell, and the virus copies go on to infect other cells. As the virus destroys helper T-cells, it steadily weakens the immune system. If enough T-cells are lost, the body then becomes highly susceptible to a range of different infections, a condition known as acquired immune deficiency syndrome (AIDS).

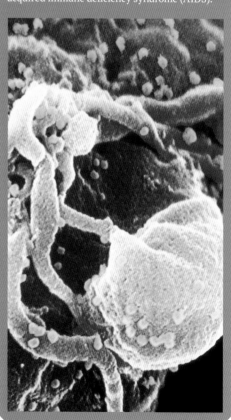

Major points of the lymph node

1. Outgoing lymph vessel
The vessel that carries filtered lymph out of the lymph node

2. Valve
A structure that prevents lymph from flowing back into the lymph node

3. Vein
Passageway for blood leaving the lymph node

4. Artery
Supply of incoming blood for the lymph node

5. Reticular fibres
Divides the lymph node into individual cells

6. Capsule
The protective, shielding fibres that surround the lymph node

7. Sinus
A channel that slows the flow of lymph, giving macrophages the opportunity to destroy any detected pathogens

8. Incoming lymph vessel
A vessel that carries lymph into the lymph node

9. Lymphocyte
The T-cells, B-cells and natural killer cells that fight infection

10. Germinal centre
This is the site of lymphocyte multiplication and maturation

11. Macrophage
Large white blood cells that engulf and destroy any detected pathogens

How old is your body?

You will make 2 million new red blood cells in the time it takes you to read this sentence

Your body contains 37.2 trillion cells. There are 86 billion neurons in your brain, 50 billion fat cells insulate your skin, and every cubic millimetre of your blood contains 4-6 million cells. But they don't live forever. Cells get old and damaged, and your body is constantly racing to replace them. Red blood cells only live for about three months; 50 million skin cells flake away every day; and sperm cells only last for three to five days. Read on to find out just how old you really are.

LENS 80+ YEARS

The lens is the part of the eye that focuses light onto the retina. It is mostly made up of fibrous tissue and water, but it contains a layer of cells that live as long as you do.

CHEEK LINING
3 hours

Studies of cheek lining cells in saliva have revealed that the lining of the mouth might renew as fast as every 2.7 hours.

STOMACH LINING
2–9 days

A thick layer of mucus protects the cells lining the stomach, but they are still replaced at least once a week.

PLATELETS
10 days

Large cells called megakaryocytes make fragments called platelets, which plug leaks in blood vessels. They only last for around ten days.

EPIDERMAL CELLS
10–30 days

There are between 18 and 23 layers of dead cells on the outside of your skin. New cells push up from below the surface every few weeks.

SPERM 3–5 days
Adult males produce fresh sperm constantly. These cells can survive for between three and five days as they search for an egg.

EGGS 50+ years
Females are born with all of the egg cells they will ever have, but they are no longer released after the menopause.

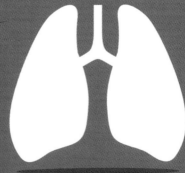

CEREBRAL NEURONS 80+ years
You might have heard that the whole body renews itself every seven years, but brain cells last as long as we do.

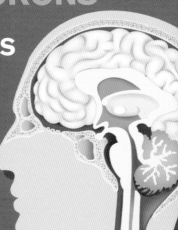

LUNG LINING 8 days
The delicate lining of the lungs is just one cell thick and lasts just over a week.

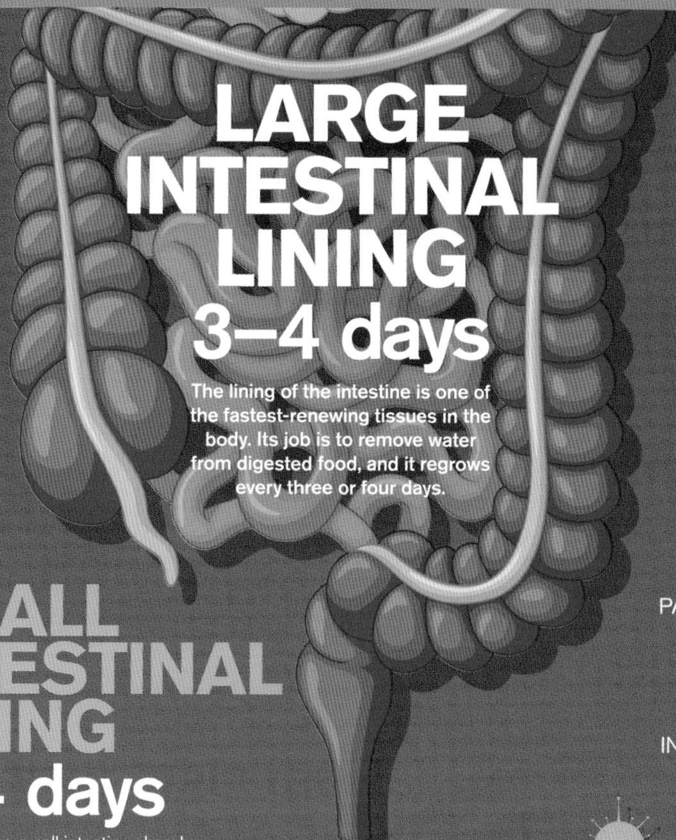

LARGE INTESTINAL LINING
3–4 days

The lining of the intestine is one of the fastest-renewing tissues in the body. Its job is to remove water from digested food, and it regrows every three or four days.

SMALL INTESTINAL LINING
2–4 days

The lining of the small intestine absorbs nutrients from digested food. It gets replaced every two to four days.

THE CELLS THAT LINE THE SMALL INTESTINE ARE SOME OF THE FASTEST-DIVIDING CELLS IN THE BODY

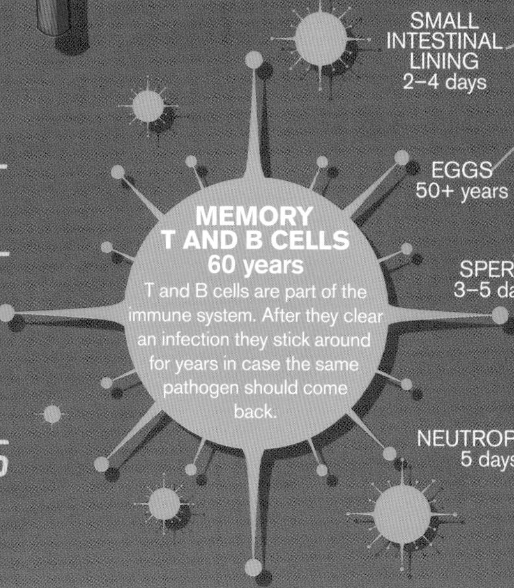

AN OVERVIEW TO YOUR BODY'S AGE

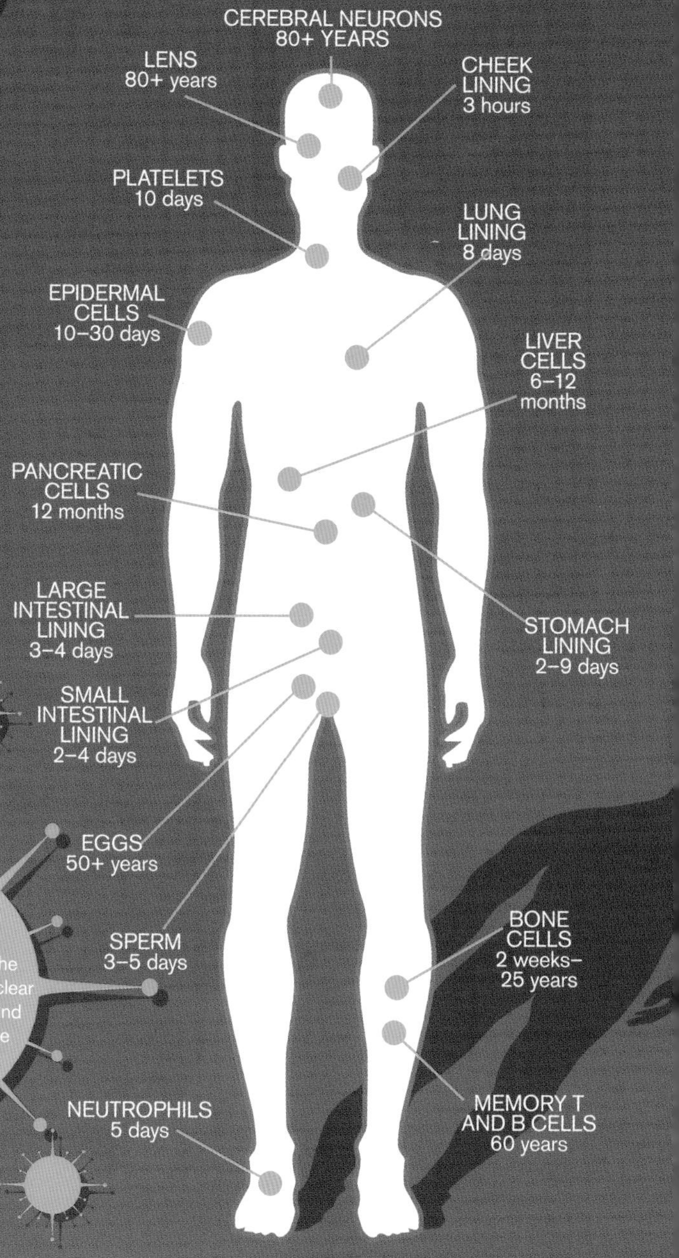

CEREBRAL NEURONS
80+ YEARS

LENS
80+ years

CHEEK LINING
3 hours

PLATELETS
10 days

LUNG LINING
8 days

EPIDERMAL CELLS
10–30 days

LIVER CELLS
6–12 months

PANCREATIC CELLS
12 months

LARGE INTESTINAL LINING
3–4 days

STOMACH LINING
2–9 days

SMALL INTESTINAL LINING
2–4 days

EGGS
50+ years

MEMORY T AND B CELLS
60 years

T and B cells are part of the immune system. After they clear an infection they stick around for years in case the same pathogen should come back.

SPERM
3–5 days

BONE CELLS
2 weeks–25 years

NEUTROPHILS
5 days

MEMORY T AND B CELLS
60 years

PANCREATIC CELLS	12 MONTHS	Beta cells in the pancreas make insulin. Their exact lifespan is still unknown, but scientists think that they live for over a year.
LIVER CELLS	6–12 MONTHS	Liver cells normally last for 200–300 days, but they can divide rapidly if needed. Remove 75 per cent of the liver and it will grow back.
BONE CELLS	2 WEEKS–25 YEARS	Bone-absorbing osteoclasts live for two weeks, bone-making osteoblasts for three months and bone-sustaining osteocytes for up to 25 years.
NEUTROPHILS	5 DAYS	White blood cells called neutrophils are first on the scene when an infection strikes. They live for less than a week.

101

Human pregnancy

Nine months of change and growth

Pregnancy is a unique period in a woman's life that brings about physical and emotional changes. When it occurs, there is an intricate change in the balance of the oestrogen and progesterone hormones, which causes the cessation of menstruation and allows the conditions in the uterus (womb) to become suitable for the growth of the fetus. The lining of the uterus, rather than being discharged, thickens and enables the development of the baby.

At first, it is a collection of embryonic cells no bigger than a pinhead. By week four the embryo forms the brain, spinal cord and heart inside the newly fluid-filled amniotic sac. Protected by this cushion of fluid, it becomes recognisably human and enters the fetal stage by the eighth week.

Many demands are put on the mother's body and she is likely to experience sickness, tiredness, lower-back pain, heartburn, increased appetite and muscle cramps, as well as the enlargement of her breasts and stretch marks. Her blood sugar levels, heart rate and breathing also increase to cope with the growing demands of the fetus.

As the date of labour approaches, the mother feels sudden contractions known as Braxton-Hicks, and the neck of her uterus begins to soften and thin out. Meanwhile, the lungs of the fetus fill with surfactant. This substance enables the lungs to soften, making them able to inflate when it takes its first breath of air in the world. Finally, chemical signals from the fetus trigger the uterus to go into labour.

"At first, it is a collection of embryonic cells no bigger than a pinhead"

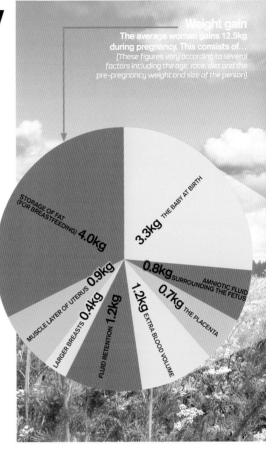

Weight gain
The average woman gains 12.5kg during pregnancy. This consists of...
(These figures vary according to several factors including the age, race, diet and the pre-pregnancy weight and size of the person)

- STORAGE OF FAT (FOR BREASTFEEDING) 4.0kg
- THE BABY AT BIRTH 3.3kg
- AMNIOTIC FLUID SURROUNDING THE FETUS 0.8kg
- THE PLACENTA 0.7kg
- EXTRA BLOOD VOLUME 1.2kg
- FLUID RETENTION 1.2kg
- LARGER BREASTS 0.4kg
- MUSCLE LAYER OF UTERUS 0.9kg

FIRST TRIMESTER (0–12 weeks)

This begins after the last menstrual period, when an egg is released and fertilised. It takes about nine weeks for the resulting embryo to develop into a fetus. During this period, the mother will be prone to sickness and mood swings due to hormonal changes.

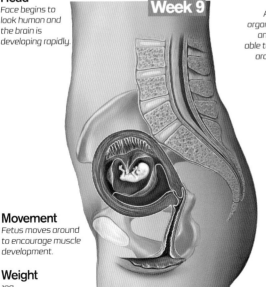

Week 9

Head
Face begins to look human and the brain is developing rapidly.

Heart
All the internal organs are formed and the heart is able to pump blood around its body.

Movement
Fetus moves around to encourage muscle development.

Weight
10g

Length
5.5cm

4 x trimester images © Science Photo Library

SECOND TRIMESTER (13–27 weeks)

The fetus grows rapidly and its organs mature. By week 20 its movements can be felt. At week 24 it can suck its thumb and hiccup, and can live independently of the mother with medical support.

Week 16

Hair and teeth
At 16 weeks, fine hair (lanugo) grows over the fetal body. By 20 weeks, teeth start forming in the jaw and hair grows.

Movement
By week 16 the eyes can move and the whole fetus makes vigorous movements.

Sound and light
The fetus will respond to light and is able to hear sounds such as the mother's voice.

Vernix
By 20 weeks, this white, waxy substance covers the skin, protecting it from the surrounding amniotic fluid.

Sweating
An increase in blood circulation causes mother to sweat more.

Weight
Week 16: 140g
Week 20: 340g

Length
Week 16: 18cm
Week 20: 25cm

The placenta

The placenta is an essential interface between the mother and fetus. When mature it is a 22cm diameter, flat oval shape with a 2.5cm bulge in the centre. The three intertwined blood vessels from the cord radiate from the centre to the edges of the placenta. Similar to tree roots, these villous structures penetrate the placenta and link to 15 to 20 lobes on the maternal surface.

The five major functions of the placenta as tasked with respiration, nutrition, excretion of waste products, bacterial protection and the production of vital hormones.

Wharton's jelly
The umbilical blood vessels are coated with this jelly-like substance and protected by a tough yet flexible outer membrane.

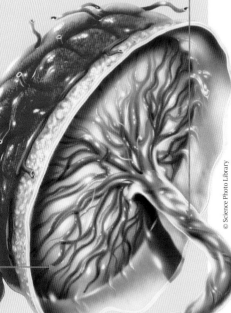

© Science Photo Library

Placenta body
Is firmly attached to the inside of the mother's uterus.

Maternal surface
Blood from the mother is absorbed and transferred to the fetal surface.

Fetal surface
Blood vessels radiate out from the umbilical cord and penetrate the placenta. The surface is covered with the thin amnion membrane.

Umbilical cord
Consists of three blood vessels. Two carry carbon dioxide and waste from the fetus, the other supplies oxygen and nutrients from the mother.

THIRD TRIMESTER (28–40 weeks)

Breathlessness
The increased size of the fetus by 24 weeks causes compression of rib cage and discomfort for mother.

Movement
By the 28th week, due to less room in uterus, the fetus will wriggle if it feels uncomfortable.

Hands
The fetus can move its hands to touch its umbilical cord at 24 weeks.

Position
By 28 weeks, the uterus has risen to a position between the navel and the breastbone.

Head
The head can move at 28 weeks and the eyes can open and see.

Now almost at full term, the fetus can recognise and respond to sounds and changes in light. Fat begins to be stored under the skin and the lungs are the very last organs to mature.

"The three intertwined blood vessels radiate from the centre to the edges of the plancenta"

Week 24

Weight
Week 24: 650g
Week 28: 1,250g

Length
Week 24: 34cm
Week 28: 38cm

Week 32

Weight
1,500g

Length
41cm

Under pressure
Pressure on the diaphragm and other organs causes indigestion and heartburn in the mother. She will find it difficult to eat a lot.

Position
Head positions itself downwards, in preparation for labour.

Sleep patterns
Fetus will sleep and wake in 20-minute cycles.

Hyperthermia vs hypothermia

What happens to the human body when the temperature is too high or too low?

The human body operates best at a temperature of around 37 degrees Celsius. We can tolerate a change of a few degrees in either direction, but any more than that and things start to go wrong.

Once body temperature drops below 35 degrees Celsius, mild hypothermia kicks in. To conserve heat, the body diverts blood away from the skin and hairs stand on end. The muscles contract and relax involuntarily, burning fuel to generate warmth. The colder the body gets, the more it starts to slow down. Nerve signals become sluggish, speech gets slurred and confusion starts to set in.

If the core temperature drops below 32 degrees Celsius, the situation becomes critical and medical attention is needed. At this point, shivering stops and the person may pass out. Below 30 degrees Celsius, the body loses its ability to warm itself up again, and this is often fatal.

The opposite of hypothermia is hyperthermia. The body has built-in mechanisms to lose heat, but sometimes it's too warm for them to work properly. If the body can't get rid of excess heat, core temperature starts to rise.

Too hot or too cold

What are the signs of hyperthermia and hypothermia?

Dizziness
The combination of dilated blood vessels and fluid loss affects blood pressure, causing dizziness.

Thirst
Water is lost to sweating, lowering the amount of fluid in the blood and triggering thirst.

Sweating
Sweating cools the skin as water evaporates, which also removes some of the excess heat.

Confusion
The cold affects cognitive function, making people feel drowsy and confused.

Altered breathing
At first, breathing speeds up, but as hypothermia progresses, both heart rate and breathing rate will slow down.

Sponging the head and neck can help to cool people down

104

Getting back to normal

Cooling someone with hyperthermia can be as simple as getting them out of their environment. This might be seeking shade on a sunny day or using fans and dehumidifiers to lower the temperature of the room. Sponging the head, neck and torso with water and offering people cold drinks can also help. For serious cases, hospital treatment can involve flushing the stomach with ice water, pumping cool air into the lungs or cooling the blood by passing it through a dialysis machine.

For hypothermia, dry blankets, warm drinks and energy-rich foods (only if they can swallow normally) can restore normal temperature, but hot baths, alcohol and rubbing the skin can be dangerous. Levels of a hormone called vasopressin drop when we are cold, leading to the production of lots of diluted urine. This can lower the amount of blood in circulation. If the blood vessels in the skin dilate too fast, the sudden drop in blood pressure can stop the heart beating.

Hypo- and hyperthermia can become very serious if they are not treated quickly. Always seek medical advice immediately if you think someone might be suffering from either condition.

Severe hyperthermia can be treated with dialysis to cool the blood

Dilated blood vessels
The blood vessels dilate, bringing warm blood to the surface of the skin.

Shivering
An automatic shivering mechanism helps to generate extra heat by contracting and relaxing the muscles.

When sweating isn't enough to lower body temperature, it can lead to dizziness and nausea. The loss of fluid triggers thirst and headaches. At the same time blood vessels dilate, bringing hot blood to the skin, but as the amount of fluid in the system drops, so too does blood pressure. This can cause dizziness and even fainting.

If the temperature climbs to over 40 degrees Celsius, molecules become misshapen and can no longer do their jobs properly, and cells start to die. Untreated, hyperthermia can lead to multiple organ failure.

Thankfully, the body has a built-in thermostat that normally keeps the temperature constant.

Pale skin
The blood vessels in the skin constrict, diverting blood to the core of the body and helping to conserve heat.

"If the core temperature drops below 32°C the situation becomes critical"

Our skin turns pale when we are cold because our blood vessels constrict

Bacteria vs virus

Which is which, and why does it even matter?

The flu virus is covered in molecules that help it to get inside cells

When you've got a sore throat, the cause doesn't always seem important. Some microscopic nasty is waging war with your immune system, it hurts, and you just want to feel better. But whether it's bacteria or a virus on the rampage is actually very important.

Bacteria are each made from just a single, primitive cell. Their insides are separated from the outside by a fatty membrane and a flexible coat of armour called the cell wall. Their genetic information is carried on loops of DNA, and these contain tiny factories called ribosomes, which use the genetic code to produce the molecules that the bacteria need to grow, divide and survive.

Viruses, on the other hand, are not technically alive. They carry genetic information containing the instructions to build more virus particles, but they don't have the equipment to make molecules themselves. To reproduce, they need to get inside a living cell and hijack its machinery, turning it into a virus factory.

Both bacteria and viruses can cause diseases, but knowing which is the culprit is critical to treating them effectively. Antibiotics can harm bacteria, but have no effect on viruses. Even your own immune system uses different tactics.

For bacteria, the immune system unleashes antibodies – projectile weapons that stick invading microbes together, slowing them down and marking them for eventual destruction. For viruses, your immune system can search for any infected cells before initiating a self-destruct sequence to dispose of anything nasty lurking inside. But some viruses are able to endure our defences, and can remain inside us indefinitely.

Head to head

Both are microscopic, but take a closer look and the differences become clear

Not alive
Viruses do not possess the tools to make their own molecules, and are missing genes vital for life.

Nucleic acid
Viruses carry genetic information; some in the form of DNA, and others in the form of RNA.

Chromosome
Bacteria carry their genetic code on a chromosome made from DNA.

Cell membrane
The membrane helps to control what goes in and out of the bacterium.

Plasmid
These small loops of DNA can be transferred between bacterial cells.

Protein coat
The virus' genetic information is stored inside a protective covering of molecules called proteins.

Envelope
Some viruses also have an outer envelope, often made from fat and protein.

Cell wall
Bacteria have a protective cell wall, which helps to maintain their structure.

Ribosome
These structures allow bacteria to make the molecules that they need to live.

© Thinkstock; Shutterstock

"For viruses to reproduce, they need to get inside a living cell and hijack its machinery, turning it into a virus factory"

Antibiotic resistance

Antibiotics attack bacteria. They work by interrupting the way that the tiny cells divide, grow and repair. However, if an infection is caused by a virus, antibiotics won't help. Viruses don't work in the same way as bacteria, so antibiotics can't help to fend them off. It might not seem like much of a problem, but every time antibiotics are used, it gives bacteria a chance to learn how to resist them. So every time a patient with a virus is given antibiotics, not only will they not get better, but bacteria lurking in their bodies will have a chance to see the drug and develop defences against it.

What is saliva?

Find out this frothy liquid's vital role in maintaining human health

Humans can produce an incredible two litres (half a gallon) of saliva each day. It is made up of 99.5 per cent water, so how is it able to perform so many important functions in our mouths? The answer lies in the remaining 0.5 per cent, which contains a host of enzymes, proteins, minerals and bacterial compounds. These ingredients help to digest food and maintain oral hygiene.

As soon as food enters the mouth, saliva's enzymes start to break it down into its simpler components, while also providing lubrication to enable even the driest snack to slide easily down the throat. Saliva is also important in oral health, as it actually helps to protect the teeth from decay and it also controls bacterial levels in the mouth in order to help reduce the overall risk of infection. Without sufficient saliva, tongue and lip movements are not as smooth, which, in extreme cases, can make it very difficult to speak.

With advanced scientific techniques and research, an individual's saliva can reveal a great deal of information. New studies have shown that a saliva test can be used to find out whether a person is at risk of a heart attack, as it contains C-reactive protein (CRP). This can be an indicator of heart disease when found at elevated levels in the blood. A saliva test is much less intrusive than a blood test and gives doctors a rough estimate of the health of a patient's heart. What's more, saliva contains your entire genetic blueprint. Even tiny amounts, equivalent to less than half a teardrop, can provide a workable DNA sample that can be frozen and thawed multiple times without breaking down.

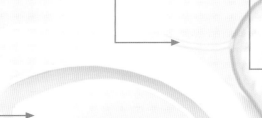

Digestive enzymes
The digestion process begins in the mouth, as saliva contains enzymes that start to break down starches and fats.

Parotid duct
The parotid duct allows saliva to move easily from the parotid gland to the mouth.

Parotid gland
The parotid glands are the largest salivary glands. They are made up of serous cells which produce thin, watery saliva.

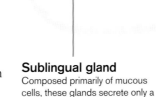

Sublingual gland
Composed primarily of mucous cells, these glands secrete only a small amount of saliva, accounting for about five per cent.

Submandibular gland
These glands produce roughly 70 per cent of your saliva. They are composed of both serous and mucous cells.

Submandibular duct
Also known as the Wharton duct, this drains saliva from both the submandibular and sublingual glands.

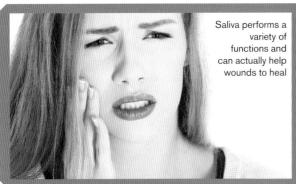

Saliva performs a variety of functions and can actually help wounds to heal

Can saliva speed up healing?

Many animals do it instinctively, but it turns out that there is a benefit to humans licking their wounds. A study found that there is a compound in human saliva, namely histatin, which can speed up the healing process. Scientists conducted an experiment using epithelial cells from a volunteer's inner cheek, creating a wound in the cells so that the healing process could be monitored. They created two dishes of cells, one that was treated with saliva and one that was left open. The scientists were astounded when after 16 hours the saliva-treated wound was almost completely closed, yet the untreated wound was still open. This demonstrated that saliva does aid the healing of at least oral wounds, something that has been suspected but unproven until this study.

Neurotransmitters and your feelings

Are our moods and emotions really just brain chemistry?

Messages are passed from one nerve cell to the next by chemical messengers called neurotransmitters. Each has a slightly different effect and by looking at what happens when neurotransmitter levels change, we are discovering that different combinations play a role in a range of complex emotions.

Acetylcholine excites the nerve cells that it touches, triggering more electrical activity. It plays a role in wakefulness, attention, learning and memory, and abnormally low levels are found in the brains of people with dementia caused by Alzheimer's disease.

Dopamine is a chemical that also excites nerve cells. It plays a vital role in the control of movement and posture, and low levels of dopamine underlie the muscle rigidity that exists in Parkinson's disease. Dopamine is also used in the brain's reward circuitry and is one of the chemicals responsible for the good feelings that are normally associated with more addictive behaviour types.

Noradrenaline is similar in structure to the hormone adrenaline and is involved in the 'fight or flight' response. In the brain, it keeps us alert and focussed. In contrast, GABA reduces the activity of the nerves that it interacts with and is thought to reduce feelings of fear or anxiety.

Serotonin is sometimes known as the 'happy hormone' and transmits signals involved in body temperature, sleep, mood and pain. People with depression have been found to have lower serotonin levels than normal, though raising serotonin levels with antidepressant medications does not always help.

There are many more neurotransmitters in the brain and other chemicals like hormones can also influence the behaviour of nerve cells. It is these interactions that are thought to underlie the huge range of human emotions.

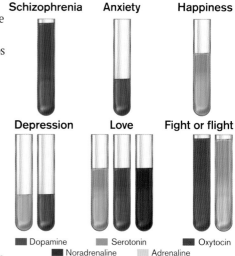

Schizophrenia Anxiety Happiness

Depression Love Fight or flight

■ Dopamine ■ Serotonin ■ Oxytocin
■ Noradrenaline ▨ Adrenaline

Different levels of neurotransmitters have been associated with different mental states

The synapse

Neurotransmitters pass messages from one nerve cell to the next

Incoming signal
Neurotransmitter release is only triggered when there is enough electrical activity in the nerve cell.

Neurotransmitters
These chemical messengers travel across the small gap - called the synaptic cleft - and stick to receptors on nearby nerve cells.

Synapse
Nerve cells communicate by releasing neurotransmitters at specialised junctions called synapses.

Part of a network
Each nerve cell makes thousands of connections to its neighbours and has its own mix of different neurotransmitters and receptors.

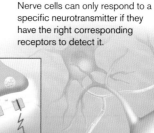

New signal
If a neighbouring nerve receives the right chemical messages it will trigger a new electrical signal.

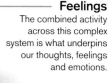

Receptor
Nerve cells can only respond to a specific neurotransmitter if they have the right corresponding receptors to detect it.

Feelings
The combined activity across this complex system is what underpins our thoughts, feelings and emotions.

© BSIP SA / Alamy; Thinkstock

Brain cells

Find out what's really going on inside your head

Your brain is an incredible thing. It is one of the most complex structures in the known universe, and for decades, scientists have been teasing it apart to find out what it's made of and how it works.

The brain is an electrical and chemical circuit, and nerve cells, or neurons, are the components. They each have a cell body, which contains their genetic code, an axon to transmit electrical impulses, and dendrites to receive them.

They are connected together at junctions known as synapses. When an impulse arrives, packets of molecules are released, passing the message on. Each neuron makes hundreds, or even thousands, of connections, producing the complicated patterns that drive human thought.

There are hundreds of different types of neuron in the brain, categorised according to their unique structure and function, and more are being discovered all the time. But they can't function on their own. They are supported by a network of glial cells – a name that literally means 'glue'.

There are three main types of glial cell. Oligodendrocytes have fatty branches, which they wrap around the conducting axons of nerve cells like the plastic coating on electrical wires. This provides insulation, preventing signals from getting crossed and speeding up their transmission along the chain.

Microglia are part of the immune system and act like an in-house cleanup crew, tracking down pathogens and clearing debris from the brain. Then there's the star-shaped astrocytes, which reach between nerve cells and blood vessels with their long, thin arms, shuttling nutrients, mopping up waste products, and even getting involved with chemical signalling.

How many cells?

It's hard to know exactly how many cells are in the brain. Individual neurons have long, thin axons and branching trees of dendrites that cross over with their neighbours, forming a tangled mass that is almost impossible to accurately examine. One of the most commonly quoted estimates is 100 billion neurons, with anywhere between three and ten times as many supporting glial cells, but the latest research suggests that these numbers are in fact wrong.

Using a new technique for counting cells, scientists have come up with a different number. Each cell has one nucleus, and they can be stained up to make it easy to tell whether they belong to a neuron or a glial cell. Rather than count them under a microscope, the researchers popped the cells open and turned them into a 'soup' so that they could be quickly counted by machine. Using this technique, they revealed that there are closer to 86 billion neurons and about the same number of glial cells – far fewer than expected.

Under the microscope

A closer look at the brain reveals a complex network of different cells

Microglia
These are specialist immune cells, helping to keep the brain healthy and free from disease.

Oligodendrocyte
These cells provide insulation, wrapping fatty membranes around the neurons to speed up their electrical signals.

Axon
This part of the neuron transmits electrical signals towards neighbouring cells.

Neuron
These are the nerve cells, responsible for transmitting and receiving the electrical and chemical signals in the brain.

Dendrite
These branching processors receive thousands of incoming signals from other neurons.

Astrocyte
These star-shaped cells support the neurons, providing nutrients, clearing waste and contributing to signalling.

This microscope image shows astrocytes grabbing on to blood vessels with their 'feet'

Synapse
Chemical signals are exchanged at these junctions, passing messages from one neuron to the next.

© Thinkstock; Shutterstock

109

How do white blood cells work?

One of the body's main defences against infection and foreign pathogens, how do these cells protect our bodies?

White blood cells, or leukocytes, are the body's primary form of defence against disease. When the body is invaded by a pathogen of any kind, the white blood cells attack in a variety of ways; some produce antibodies, while others surround and ultimately devour the pathogens whole.

In total, there are five types of white blood cell (WBC), and each cell works in a different way to fight a variety of threats. These five cells sit in two groupings: the granulocytes and the agranulocytes. The groups are determined based on whether a cell has 'granules' in the cytoplasm. These granules are digestive enzymes that help break down pathogens. Neutrophils, eosinophils and basophils are all granulocytes, the enzymes in which also give them a distinct colouration which the agranulocytes do not have.

As the most common WBC, neutrophils make up between 55 and 70 per cent of the white blood cells in a normal healthy individual, with the other four types (eosinophils, basophils, monocytes and lymphocytes) making up the rest. Neutrophils are the primary responders to infection, actively moving to the site of infection following a call from mast cells after a pathogen is initially discovered. They consume bacteria and fungus that has broken through the body's barriers in a process called phagocytosis.

Lymphocytes – the second-most common kind of leukocyte – possess three types of defence cells: B cells, T cells and natural killer cells. B cells release antibodies and activate T cells, while T cells attack diseases such as viruses and tumours when directed, and regulatory T cells ensure the immune system returns to normal after an attack. Natural killer cells, meanwhile, aid T cell response by also attacking virus-infected and tumour cells, which lack a marker known as MHC.

The remaining types of leukocyte release chemicals such as histamine, preparing the body for future infection, as well as attacking other causes of illness like parasites.

"Natural killer cells aid T cell response by also attacking virus-infected and tumour cells"

Types of leukocyte

Different kinds of WBC have different roles, which complement one another to defend the body

Lymphocyte
These release antibodies as well as attack virus and tumour cells through three differing types of cell. As a group, they are some of the longest lived of the white blood cells with the memory cells surviving for years to allow the body to defend itself if repeat attacks occur.

Monocyte
Monocytes help prepare us for another infection by presenting pathogens to the body, so that antibodies can be created. Later in their life, monocytes move from the bloodstream into tissue, and then evolve into macrophages which can conduct phagocytosis.

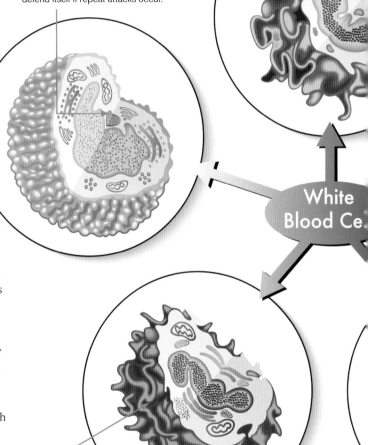

White Blood Cell

Eosinophil
Eosinophils are the white blood cells that primarily deal with parasitic infections. They also have a role in allergic reactions. They make up a fairly small percentage of the total white blood cells in our body – about 2.3 per cent.

White blood cells at work

The body has various outer defences against infection, including the external barrier of the skin, but what happens when this is breached?

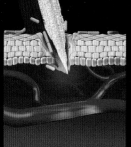

Skin breach
A foreign object breaks through the skin, introducing bacteria (shown in green) into the body.

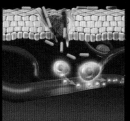

Mast cells
Mast cells release cytokines and then WBCs are called into action to ensure the infection does not spread.

WBCs arrive
Macrophages move to the site via the bloodstream to start defending against invading bacteria.

Macrophages consume bacteria
Bacteria are absorbed into cytoplasm and broken down by the macrophages.

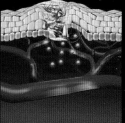

Healing
Following removal of the bacteria, the body will start to heal the break in the skin to prevent further infection.

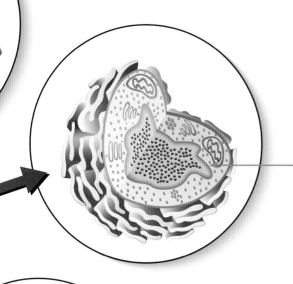

Basophil
Basophils are involved in allergic response via releasing histamine and heparin into the bloodstream. Their functions are not fully known and they only account for 0.4 per cent of the body's white blood cells. Their granules appear blue when viewed under a microscope.

Neutrophil
Neutrophils are the most common of the leukocytes. They have a short life span so need to be constantly produced by the bone marrow. Their granules appear pink and the cell has multi-lobed nuclei which make them easily differentiated from other types of white blood cell.

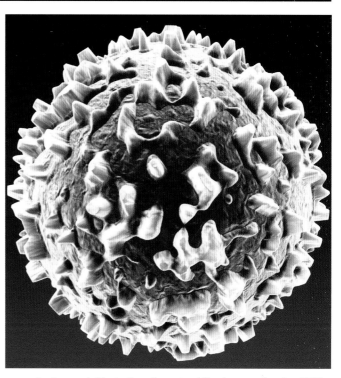

A microscopic illustration of a neutrophil – the most abundant WBC

A faulty immune system

If the immune system stops working properly, we are at risk of becoming ill. However, another problem is if the immune system actually goes into overdrive and starts attacking the individual's own cells, mistaking them for pathogens. There are a large number of autoimmune ailments seen across the world, such as Crohn's disease, psoriasis, lupus and some cases of arthritis, as well as a large number of diseases that are suspected to have autoimmune roots.

We can often treat these conditions with immunosuppressants, which deactivate elements of the immune system to stop the body attacking itself. However, there are drawbacks with this treatment as, if the person exposes themselves to another pathogen, they would not have the normal white blood cell response. Consequently, the individual is less likely to be able to fight normally low-risk infections and, depending on the pathogen, they can even be fatal.

© SPL; Thinkstock

GENETICS

From inheritance to genetic diseases, what secrets are hidden in our genes and how do they determine who we are?

How is our genetic code stored?

Genenes define who we are. They are the basic unit of heredity, each containing a coded set of instructions to make a protein. Humans have an estimated 20,500 genes, varying in length from a few hundred to more than 2 million base pairs. They affect all aspects of our physiology, providing the code that determines our physical appearance, the biochemical reactions that occur inside our cells and even, many argue, our personalities.

Every individual has two copies of every gene – one inherited from each parent. Within the population there are several alleles of each gene – that is, different forms of the same code, with a number of minor alterations in the sequence. These alleles perform the same underlying function, but it is the subtle differences that make each of us unique.

Inside each of our cells (except red blood cells) is a nucleus, the core which contains our genetic information: deoxyribonucleic acid (DNA). DNA is a four-letter code made up of bases: adenine (A), guanine (G), cytosine (C) and thymine (T). As molecular biologist Francis Crick once put it, "DNA makes RNA, RNA makes protein and proteins make us." Our genes are stored in groups of several thousand on 23 pairs of chromosomes in the nucleus, so when a cell needs to use one particular gene, it makes a temporary copy of the sequence in the form of ribonucleic acid (RNA). This copy contains all of the information required to make a protein – the building blocks of the human body.

Genetic information is coded into DNA using just four nucleobases: A, C, G and T

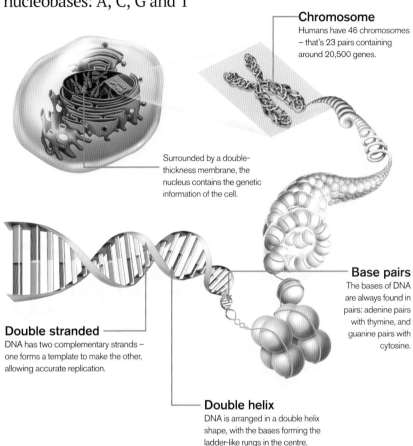

Chromosome
Humans have 46 chromosomes – that's 23 pairs containing around 20,500 genes.

Surrounded by a double-thickness membrane, the nucleus contains the genetic information of the cell.

Double stranded
DNA has two complementary strands – one forms a template to make the other, allowing accurate replication.

Base pairs
The bases of DNA are always found in pairs: adenine pairs with thymine, and guanine pairs with cytosine.

Double helix
DNA is arranged in a double helix shape, with the bases forming the ladder-like rungs in the centre.

DNA's chemical structure

We put deoxyribonucleic acid under the microscope

Nucleotide
DNA is a polymer made up of building blocks called nucleotides.

Phosphate
Phosphate groups link the sugars of adjacent nucleotides together, forming a phosphate backbone.

Hydrogen bond
Two bases interact with each other by hydrogen bonds (weak electrostatic interactions that hold the strands of DNA together).

Sugar
Each base is attached to a five-carbon sugar called deoxyribose.

Nucleobase
Each nucleotide contains a base, which can be one of four: adenine (A), thymine (T), guanine (G) or cytosine (C).

T A

The Human Genome Project aimed to map the entire human genome; this map is effectively a blueprint for making a human. Using the information hidden within our genetic code, scientists have been able to identify genes that contribute to various diseases. By logging common genetic variation in the human population, researchers have actually been able to identify over 1,800 disease-associated genes, affecting illnesses ranging from breast cancer to Alzheimer's. The underlying genetic influences that affect complex diseases such as heart disease are still not yet fully understood, but having the genome available to study is making the task of identifying the genetic risk factors much easier.

Interestingly, the Human Genome Project discovered we have far fewer genes than first predicted; in fact, only two per cent of our genome codes for proteins. The remainder of the DNA is known as 'non-coding' and serves other functions. In many human genes are non-coding regions called introns, and between genes there is intergenic DNA. One proposed function is that these sequences act as a buffer to protect the important genetic information from mutation. Other

non-coding DNA acts as switches, which helps the cell to turn genes on and off at the right times.

Genetic mutations are the source of variation in all organisms. Most genetic mutation occurs as the DNA is being copied, when cells prepare to divide. The molecular machinery responsible for duplicating DNA is prone to errors, and often makes mistakes, resulting in changes to the DNA sequence. These can be as simple as accidentally substituting one base for another (eg A for G), or can be much larger errors, like adding or deleting bases. Cells have repair machinery to correct errors as they occur, and even to kill the cell if it makes a big mistake, but despite this some errors still slip through.

Throughout your life you will acquire many cell mutations. Many of these are harmless, either occurring in non-coding regions of DNA, or changing the gene so nominally that the protein is virtually unaffected. However, some mutations do lead to disease.

If mutations are introduced into the sperm and egg cells they can be passed on to the next generation. However, not all mutations are bad, and this process of randomly introduced changes in the DNA

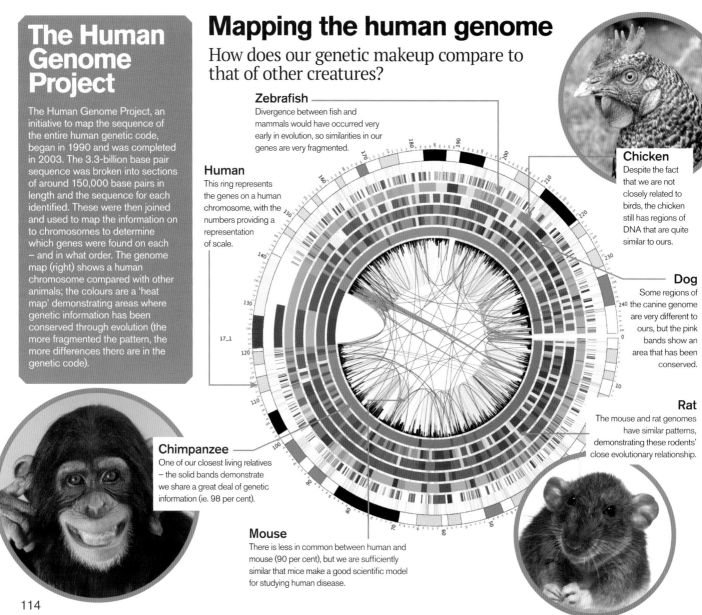

The Human Genome Project

The Human Genome Project, an initiative to map the sequence of the entire human genetic code, began in 1990 and was completed in 2003. The 3.3-billion base pair sequence was broken into sections of around 150,000 base pairs in length and the sequence for each identified. These were then joined and used to map the information on to chromosomes to determine which genes were found on each – and in what order. The genome map (right) shows a human chromosome compared with other animals; the colours are a 'heat map' demonstrating areas where genetic information has been conserved through evolution (the more fragmented the pattern, the more differences there are in the genetic code).

Mapping the human genome

How does our genetic makeup compare to that of other creatures?

Zebrafish
Divergence between fish and mammals would have occurred very early in evolution, so similarities in our genes are very fragmented.

Human
This ring represents the genes on a human chromosome, with the numbers providing a representation of scale.

Chicken
Despite the fact that we are not closely related to birds, the chicken still has regions of DNA that are quite similar to ours.

Dog
Some regions of the canine genome are very different to ours, but the pink bands show an area that has been conserved.

Rat
The mouse and rat genomes have similar patterns, demonstrating these rodents' close evolutionary relationship.

Chimpanzee
One of our closest living relatives – the solid bands demonstrate we share a great deal of genetic information (ie. 98 per cent).

Mouse
There is less in common between human and mouse (90 per cent), but we are sufficiently similar that mice make a good scientific model for studying human disease.

sequence provides the biological underpinning that supports Darwin's theory of evolution. This is most easily observed in animals. Take, for example, the peppered moth. Before the Industrial Revolution the majority of these moths had white wings, enabling them to hide against light-coloured trees and lichens. A minority had a mutant gene, which gave them black wings; this made them an easy target for predators. When factories began to cover the trees in soot, the light-coloured moths struggled to hide themselves against the darker environment, so black moths flourished. They survived much longer, enabling them to pass on their mutation to their offspring and altering the gene pool.

It is easy to see how a genetic change like the one that occurred in the peppered moth could give an advantage to a species, but what about genetic diseases? Even these can work to our advantage. A good example is sickle cell anaemia – a genetic disorder that's quite common in the African population.

A single nucleotide mutation causes haemoglobin, the protein involved in binding oxygen in red blood cells, to misfold. Instead of

"Before the Industrial Revolution the majority had white wings"

forming its proper shape, the haemoglobin clumps together, causing red blood cells to deform. They then have trouble fitting through narrow capillaries and often become damaged or destroyed. However, this genetic mutation persists in the population because it has a protective effect against malaria. The malaria parasite spends part of its life cycle inside red blood cells and, when sickle cells rupture, it prevents the parasite from reproducing. Individuals with one copy of the sickle cell gene and one copy of the healthy haemoglobin gene have few symptoms of sickle cell anaemia, but are protected from malaria too, allowing them to pass the gene on to their children.

Using genetics to convict criminals

Forensic scientists can use traces of DNA to identify individuals involved in criminal activity. Only about 0.1 per cent of the genome differs between individuals, so rather than sequencing the entire genome, scientists take 13 DNA regions that are known to vary between different people in order to create a 'DNA fingerprint'. In each of these regions there are two to 13 nucleotides in a repeating pattern hundreds of bases long – the length varies between individuals. Small pieces of DNA – referred to as probes – are used to identify these repeats and the length of each is determined by a technique called polymerase chain reaction (PCR). The odds that two people will have exactly the same 13-region profile is thought to be one in a billion or even less, so if all 13 regions are found to be a match then scientists can be fairly confident that they can tie a person to a crime scene.

Why do we look like our parents?

It's a common misconception that we inherit entire features from our parents – eg "You have your father's eyes." Actually inheritance is much more complicated – several genes work together to create traits in physical appearance; even eye colour isn't just down to one gene that codes for 'blue', 'brown' or 'green', etc. The combinations of genes from both of our parents create a mixture of their traits. However, there are some examples of single genes that do dictate an obvious physical characteristic all on their own. These are known as Mendelian traits, after the scientist Gregor Mendel who studied genetic inheritance in peas in the 1800s. One such trait is albinism – the absence of pigment in the skin, hair and eyes due to a defect in the protein that makes melanin.

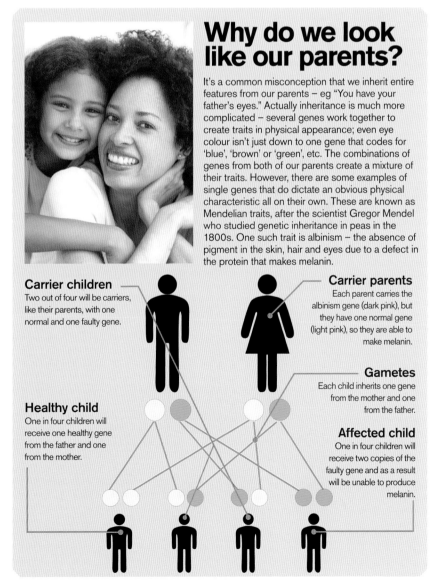

Carrier children
Two out of four will be carriers, like their parents, with one normal and one faulty gene.

Carrier parents
Each parent carries the albinism gene (dark pink), but they have one normal gene (light pink), so they are able to make melanin.

Healthy child
One in four children will receive one healthy gene from the father and one from the mother.

Gametes
Each child inherits one gene from the mother and one from the father.

Affected child
One in four children will receive two copies of the faulty gene and as a result will be unable to produce melanin.

Genetics is a complex and rapidly evolving field and more information about the function of DNA is being discovered all the time. It is now known that environmental influences can alter the way that DNA is packaged in the cell, restricting access to some genes and altering protein expression patterns. Known as epigenetics, these modifications do not actually alter the underlying DNA sequence, but regulate how it is accessed and used by the cell. Epigenetic changes can be passed on from one cell to its offspring, and provide an additional mechanism by which genetic information can be modified across generations.

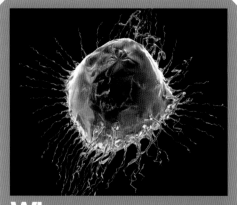

When our genes go wrong…

Cancer is not just the result of one or two genetic mutations – in fact, it takes a whole series of mistakes for a tumour to form. Cells contain oncogenes and tumour suppressor genes, whose healthy function is to tell the cell when it should and should not divide. If these become damaged, the cell cannot switch off its cell division programme and it will keep making copies of itself indefinitely. Each time a cell divides there is a risk that it will make a mistake when copying its DNA, and gradually the cell makes more and more errors, accumulating mutations that allow the tumour to progress into malignant cancer.

"Environmental influences can alter the way that DNA is packaged"

Repairing faulty genes

We reveal how donated cells can be used to mend any damaged genes within the human body

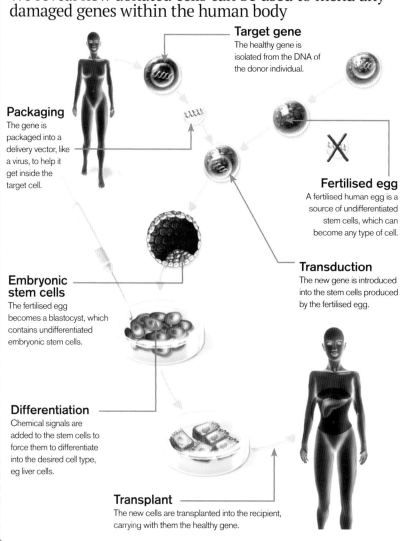

Target gene
The healthy gene is isolated from the DNA of the donor individual.

Packaging
The gene is packaged into a delivery vector, like a virus, to help it get inside the target cell.

Fertilised egg
A fertilised human egg is a source of undifferentiated stem cells, which can become any type of cell.

Transduction
The new gene is introduced into the stem cells produced by the fertilised egg.

Embryonic stem cells
The fertilised egg becomes a blastocyst, which contains undifferentiated embryonic stem cells.

Differentiation
Chemical signals are added to the stem cells to force them to differentiate into the desired cell type, eg liver cells.

Transplant
The new cells are transplanted into the recipient, carrying with them the healthy gene.

How tumours develop

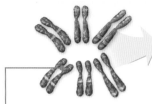

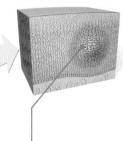

Tumour-associated genes
Genes normally involved in regulating cell behaviour can go on to cause cancer if they become mutated.

Mutagens
Environmental factors, or mutagens – such as radiation and chemicals – can cause damage to the DNA, leading to mutations in key genes.

Localised
Cancer usually starts with just one or a few mutated cells; these begin to divide uncontrollably in their local area creating a tumour.

Invasion
As the tumour grows in size it starts to invade the surrounding area, taking over neighbouring tissues.

Metastasis
Further mutations allow cells of the tumour to break free and enter the bloodstream. From here they can be distributed throughout the body.

What is anxiety?

How our brains trigger a fight or flight response

Some people who suffer anxiety find it hard to leave the house

Anxiety affects a huge number of people and can be so severe that it stops many sufferers from leaving their homes or doing their jobs. In the US, over 40 million people aged 18 or over endure an anxiety related disorder, while in the UK one in 20 people are affected. Some researchers believe that modern day technology has influenced the rise of anxiety related conditions; we are constantly on high alert with texts, emails, social media and news updates.

Anxiety is a natural human response that serves a purpose. From a biological point of view, it functions to create a heightened sense of awareness, preparing us for potential threats. In a way, it's nature's panic button. When we become anxious our fight or flight response is triggered, flooding our bodies with epinephrine (adrenaline), norepinephrine (noradrenaline) and cortisol, which help increase your reflexes and reaction speed. Your body prepares itself to deal with potential danger by increasing the heart rate, pumping more blood to the muscles and by getting the lungs to hyperventilate.

At the same time, the brain stops thinking about pleasurable things, making sure that all of its focus is on identifying potential threats. In extreme cases, the body will respond to anxiety by emptying the digestive tract by any means necessary, as this ensures that no energy is wasted on digestion.

How your brain reacts

The body's primal response to danger can be triggered by non-threatening situations

Thalamus
Visual and auditory stimuli are first processed by the thalamus which filters the incoming information and sends it to the areas where it can be interpreted.

Two paths
A startling signal such as a sudden loud noise will be sent from the thalamus via two paths: one travels directly to the amygdala - where it can quickly initiate the fear response - and the other passes through the cortex to be processed more thoroughly.

Stria terminalis
The bed nucleus of the stria terminalis (BNST) is responsible for maintaining fear once this emotion has been stimulated by the amygdala, leading to longer-term feelings of anxiety.

Amygdala
This is where the fear response is triggered. The amygdala can quickly put your body on high alert, and research suggests that if this area of the brain is overactive, it may cause an anxiety disorder.

Cortex
Once the amygdala and hippocampus have received a stimulus, the cortex's role is to find out what's caused the fear response. Once the perceived danger is over, a section of the prefrontal cortex signals the amygdala to cease its activity. It is vital to turning off anxiety.

Locus caeruleus
This area of the brain stem is triggered by the amygdala to initiate the physiological responses to anxiety or stress, such as an increase in heart rate and pupil dilation.

Hippocampus
The hippocampus is the brain's memory centre, responsible for encoding any threatening events that we experience in life into long-term memories.

© Alamy; Thinkstock

Inside the circulatory system

Arteries and veins form the plumbing system that carries blood around the body. Find out more about the circular journey it takes...

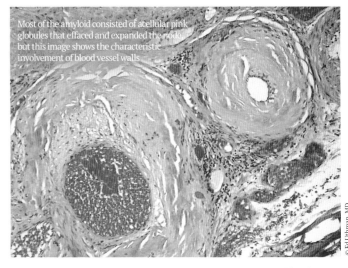

Most of the amyloid consisted of acellular pink globules that effaced and expanded the node, but this image shows the characteristic involvement of blood vessel walls

© Ed Uthman, MD

The network of blood vessels in the human body must cope with different volumes of blood travelling at different pressures. These blood vessels come in a multitude of different sizes and shapes, from the large, elastic aorta down to very tiny, one-cell-thick capillaries.

Blood is the ultimate multitasker. It carries oxygen for various tissues to use, nutrients to provide energy, removes waste products and even helps you warm up or cool down. It also carries vital clotting factors which stop us bleeding. Blood comes in just two varieties; oxygen-rich (oxygenated) blood is what the body uses for energy, and is bright red. After it has been used, this oxygen-depleted (deoxygenated) blood is returned for recycling and is actually dark red (not blue, as is often thought).

Blood is carried in vessels, of which there are two main different types – arteries and veins. Arteries carry blood away from the heart and deal with high pressures, and so have strong elastic walls. Veins carry blood back towards the heart and deal with lower pressures, so have thinner walls. Tiny capillaries connect arteries and veins together, like small back-roads connecting motorways to dual carriageways.

Arteries and veins are constructed differently to cope with the varying pressures, but work in tandem to ensure that the blood reaches its final destination. However, sometimes things go wrong, lead to certain medical problems: varicose veins from failing valves; deep vein thrombosis from blood clots blocking the deep venous system; heart attacks from blocked arteries; and lastly life-threatening aneurysms from weak artery walls.

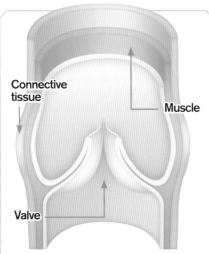

Connective tissue

Muscle

Valve

How do veins work?

Veins carry low pressure blood. They contain numerous one-way valves which stop backwards flow of blood, which can occur when pressure falls in-between heartbeats. Blood flows through these valves towards the heart but cannot pass back through them in the other direction. Valves can fail over time, especially in the legs. This leads to saggy, unsightly veins, known as varicose veins.

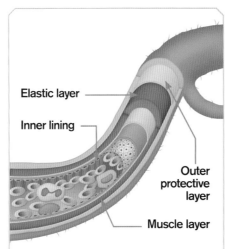

Elastic layer

Inner lining

Outer protective layer

Muscle layer

Arteries – under pressure!

Arteries cope with all of the pressure generated by the heart and deliver oxygen-rich blood to where it needs to be 24 hours a day. The walls of arteries contain elastic muscles, which allow them to stretch and contract to cope with the wide changes in pressure which is generated from the heart. Since the pressure is high, valves are unnecessary, unlike the low-pressure venous system.

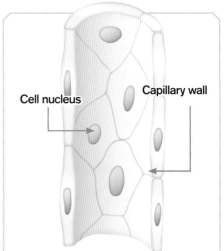

Cell nucleus

Capillary wall

Connecting it all together

Capillaries are the tiny vessels which connect small arteries and veins together. Their walls are only one cell thick, so this is the perfect place to trade substances with surrounding tissues. Red blood cells within these capillaries trade water, oxygen, carbon dioxide, nutrients, waste and even heat. Because these vessels are only one cell wide, the cells have to line up to pass through.

A game of two halves

In human beings, the heart is a double pump, meaning that there are two sides to the circulatory system. The left side of the heart pumps oxygen and nutrient-rich blood to the brain, vital organs and other body tissues (the systemic circulation). The right side of the heart pumps deoxygenated blood towards the lungs, so it can pick up new oxygen molecules to be used again (the pulmonary circulation).

"Plasma carries all of the different types of cells"

Arteries
All arteries carry blood away from the heart. They carry oxygenated blood, except for the pulmonary artery, which carries deoxygenated blood to the lungs.

Lungs
In the lungs, carbon dioxide is expelled from the body and is swapped for fresh oxygen from the air. This oxygen-rich blood takes on a bright red colour.

Aorta
The aorta is an artery which carries oxygenated blood to the body; it is the largest blood vessel in the body and copes with the highest pressure blood.

What's in blood?
It's actually only the iron in red blood cells which make blood red – if you take these cells away then what you will be left with is a watery yellowish solution that is called plasma. Plasma carries all of the various different types of cells and also contains sugars, fats, proteins and salts. The main types of cell are red blood cells (which are formed from iron and haemoglobin, which carries oxygen around the body), white blood cells (which fight infection from bacteria, viruses and fungi) and finally platelets (which are actually tiny cell fragments which stop bleeding by forming clots at the sites of any damage).

The left side
The left side of the heart pumps oxygenated blood for the body to use. It pumps directly into arteries towards the brain and other body tissues.

Veins
All veins carry blood to the heart. They carry deoxygenated blood, except for the pulmonary vein, which carries oxygenated blood back to the heart.

The right side
The right side of the heart pumps deoxygenated blood to the lungs, where blood exchanges carbon dioxide for fresh oxygen.

Capillaries
Tiny capillaries connect arteries and veins together. They allow exchange of oxygen, nutrients and waste in the body's organs and tissues.

Blood vessels
Different shapes and sizes

Capillary sphincter muscles
These tiny muscles can open and close, which can decrease or increase blood flow through a capillary bed. When muscles exercise, these muscles relax and blood flow into the muscle increases.

Capillary bed
This is the capillary network that connects the two systems. Here, exchange of various substances occurs with surrounding tissues, through the one-cell thick walls.

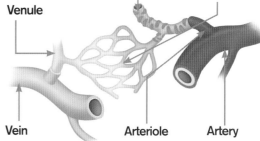

Venule · Vein · Arteriole · Artery

HEAD AND ARMS · LUNG · LUNG · HEART · LIVER · KIDNEY · TRUNK AND LEGS

How your blood works

The science behind the miraculous fluid that feeds, heals and fights for your life

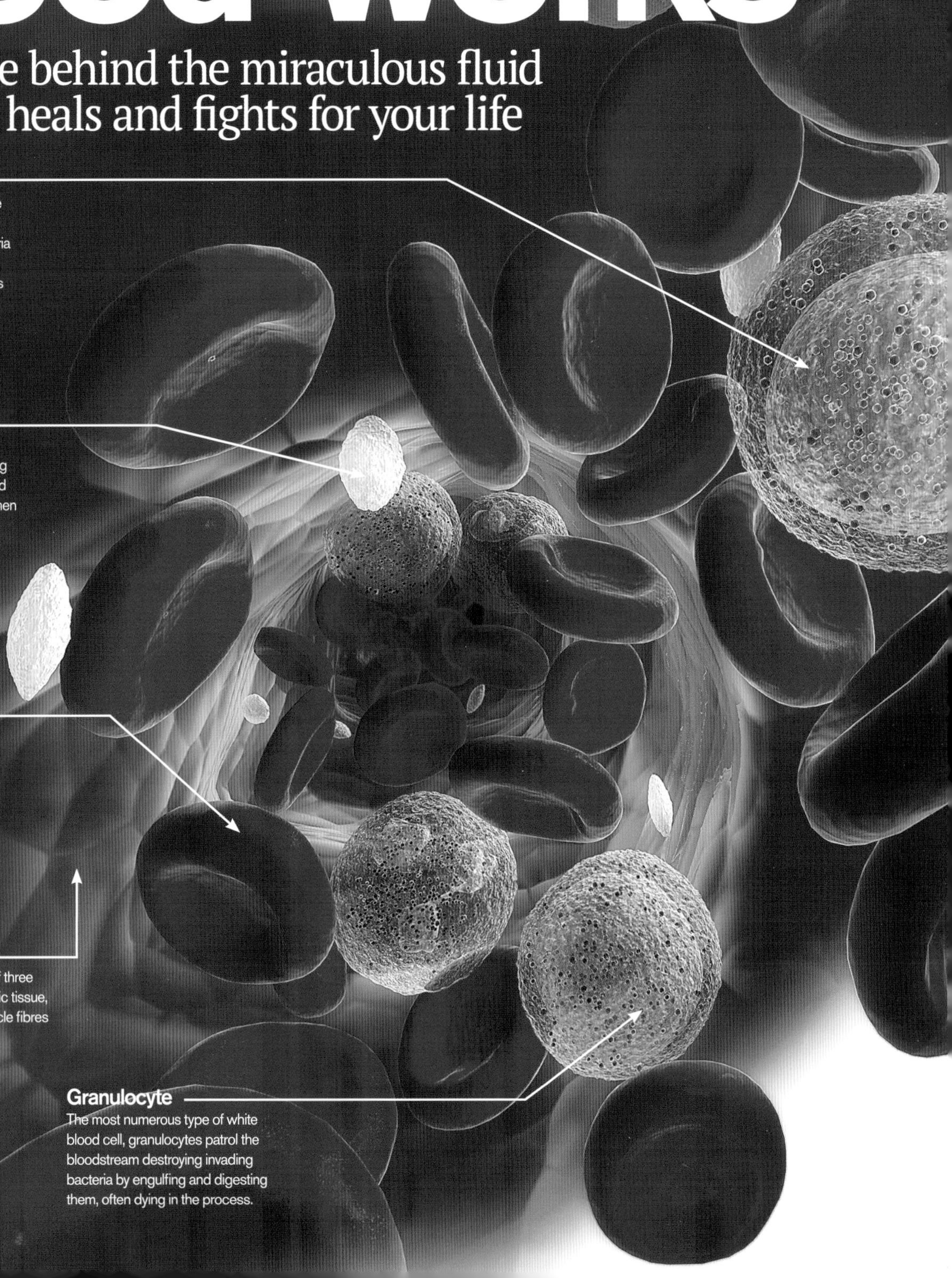

White blood cells

White blood cells, or leukocytes, are the immune system's best weapon, searching out and destroying bacteria and producing antibodies against viruses. There are five different types of white blood cells, all with distinct functions.

Platelet

When activated, these sticky cell fragments are essential to the clotting process. Platelets adhere to a wound opening to stem the flow of blood, then they team with a protein called fibrinogen to weave tiny threads that trap blood cells.

Red blood cell

Known as erythrocytes, red blood cells are the body's delivery service, shuttling oxygen from the lungs to living cells throughout the body and returning carbon dioxide as waste.

Blood vessel wall

Arteries and veins are composed of three tissue layers, a combination of elastic tissue, connective tissue and smooth muscle fibres that contract under signals from the sympathetic nervous system.

Granulocyte

The most numerous type of white blood cell, granulocytes patrol the bloodstream destroying invading bacteria by engulfing and digesting them, often dying in the process.

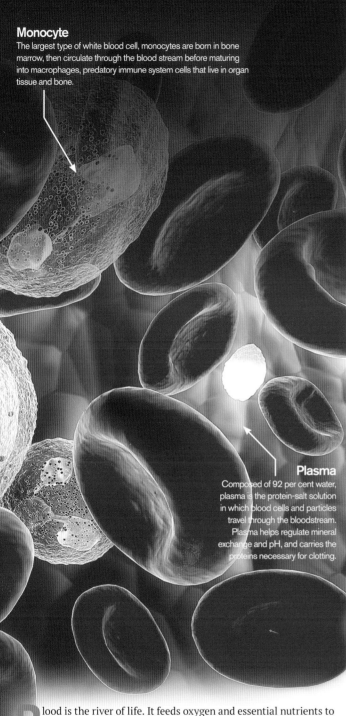

Monocyte
The largest type of white blood cell, monocytes are born in bone marrow, then circulate through the blood stream before maturing into macrophages, predatory immune system cells that live in organ tissue and bone.

Plasma
Composed of 92 per cent water, plasma is the protein-salt solution in which blood cells and particles travel through the bloodstream. Plasma helps regulate mineral exchange and pH, and carries the proteins necessary for clotting.

Components of blood

Blood is a mix of solids and liquids, a blend of highly specialised cells and particles suspended in a protein-rich fluid called plasma. Red blood cells dominate the mix, carrying oxygen to living tissue and returning carbon dioxide to the lungs. For every 600 red blood cells, there is a single white blood cell, of which there are five different kinds. Cell fragments called platelets use their irregular surface to cling to vessel walls and initiate the clotting process.

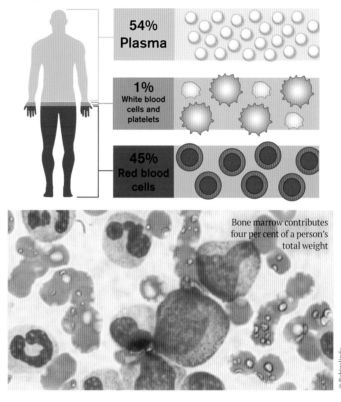

54% Plasma

1% White blood cells and platelets

45% Red blood cells

Bone marrow contributes four per cent of a person's total weight

© Bobigalindo

"Red blood cells are so numerous because they perform the most essential function of blood"

Blood is the river of life. It feeds oxygen and essential nutrients to living cells and carries away waste. It transports the foot soldiers of the immune system, white blood cells, which seek out and destroy invading bacteria and parasites. And it then speeds platelets to the site of injury or tissue damage, triggering the body's miraculous process of self-repair.

Blood looks like a thick, homogenous fluid, but it's actually more like a watery current of plasma – a straw-coloured, protein-rich fluid – carrying billions of microscopic solids consisting of red blood cells, white blood cells and cell fragments that are called platelets. The distribution is far from equal. Over half of our blood is actually just plasma, 45 per cent is red blood cells and a tiny fragment, less than one per cent, is composed of white blood cells and platelets.

Red blood cells are so numerous because they perform the most essential function of blood, which is to deliver oxygen to every cell in the body and carry away carbon dioxide. As an adult, all of your red blood cells are produced in red bone marrow, the spongy tissue in the bulbous ends of long bones and at the centre of flat bones like hips and ribs. In the marrow, red blood cells start out as undifferentiated stem cells called hemocytoblasts. If the body detects a drop in oxygen carrying capacity, a hormone is released from the kidneys that triggers the stem cells to become red blood cells. Because red blood cells only live 120 days, the supply is continuously replenished; roughly 2 million red blood cells every second.

A mature red blood cell has no nucleus, it is spit out during the final stages of the two-day development before taking on the shape of a concave, doughnut-like disc. Red blood cells are mostly water, but 97 per cent of their solid matter is haemoglobin, a complex protein that carries four atoms of iron. Those iron atoms have the ability to form loose, reversible bonds with both

Life cycle of red blood cells

6. Reuse and recycle
As for the globin and other cellular membranes, everything is converted back into basic amino acids, some of which will be used to create more red blood cells.

Waste product of blood cell

1. Born in the bones
When the body detects a low oxygen carrying capacity, hormones released from the kidney trigger the production of new red blood cells inside red bone marrow.

2. One life to live
Mature red blood cells, also known as erythrocytes, are stripped of their nucleus in the final stages of development, meaning they can't divide to replicate.

Waste excreted from body

Every second, roughly 2 million red blood cells decay and die. The body is keenly sensitive to blood hypoxia – reduced oxygen carrying capacity – and triggers the kidney to release a hormone called erythropoietin. The hormone stimulates the production of more red blood cells in bone marrow. Red blood cells enter the bloodstream and circulate for 120 days before they begin to degenerate and are swallowed up by roving macrophages in the liver, spleen and lymph nodes. The macrophages extract iron from the haemoglobin in the red blood cells and release it back into the bloodstream, where it binds to a protein that carries it back to the bone marrow, ready to be recycled in fresh red blood cells.

5. Iron ions
In the belly of Kupffer cells, haemoglobin molecules are split into heme and globin. Heme is broken down further into bile and iron ions, some of which are carried back and stored in bone marrow.

4. Ingestion
Specialised white blood cells in the liver and spleen called Kupffer cells prey on dying red blood cells, ingesting them whole and breaking them down into reusable components.

3. In circulation
Red blood cells pass from the bone marrow into the bloodstream, where they circulate for around 120 days.

oxygen and carbon dioxide – think of them as weak magnets – making red blood cells such an effective transport system for all of the respiratory gasses. Haemoglobin, which turns bright red when oxygenated, is what gives blood its characteristic crimson colour.

To provide oxygen to every living cell, red blood cells must be pumped through the body's circulatory system. The right side of the heart pumps CO_2-heavy blood into the lungs, where it releases its waste gasses and picks up oxygen. The left side of the heart then automatically pumps all of the freshly oxygenated blood out into the body through a system of various arteries and capillaries, some are even as narrow as a single cell. As the red blood cells release their oxygen, they pick up carbon dioxide molecules, then they course through the veins back toward the heart, where they are pumped back into the lungs to 'exhale' the excess CO_2 and collect some more precious O_2.

White blood cells are actually greatly outnumbered by red blood cells, but they are critical to the function of the immune system. Most white blood cells are also produced in red bone marrow, but white blood cells – unlike red blood cells – come in five different varieties, each with its own specialised immune function. The first three varieties of blood cells, are called granulocytes, engulf and digest bacteria and parasites, and play a role in allergic reactions. Lymphocytes, another type of white blood cell,

produce anti-bodies that build up our immunity to repeat intruders. And monocytes, the largest of the white blood cells, enter organ tissue and become macrophages, microbes that ingest bad bacteria and then help break down dead red blood cells into reusable parts.

Platelets aren't cells at all, they are actually tiny fragments from much larger stem cells found in bone marrow. In their resting state, they look like smooth oval plates, but when activated to form a clot they take on an irregular form with many protruding arms called pseudopods. This shape is what helps them to be able to stick not only to the blood vessel walls but also to each other, forming a physical barrier around wound sites. With the help of proteins and clotting factors that are found inside plasma, platelets weave a mesh of fibrin that stems blood loss and triggers the formation of new collagen and skin cells.

But even these three functions of blood – oxygen supplier, immune system defender and wound healer – only begin to scratch the surface of the critical role of blood in each and every bodily process. When blood circulates through the small intestine, it absorbs sugars from digested food, which are transported to the liver to be stored as energy. When blood passes through the kidneys, it is scrubbed of excess urea and salts, waste that will leave the body as urine. The proteins transport vitamins, hormones, enzymes, sugar and electrolytes.

Haemophilia

This rare genetic blood disorder severely inhibits the clotting mechanism of blood, causing excessive bleeding, internal bruising and joint problems. Platelets are essential to the clotting and healing process, producing threads of fibrin with help from proteins in the bloodstream called clotting factors. People who suffer from haemophilia – almost exclusively males – are missing one of those clotting factors, making it difficult to seal off blood vessels after even minor injuries.

Thalassemia

Another rare blood disorder affecting 100,000 newborns worldwide each year, thalassemia inhibits the production of haemoglobin, leading to severe anaemia. People who are born with the most serious form of the disease, also called Cooley's anaemia, suffer from enlarged hearts, livers and spleens, and brittle bones. The most effective treatment is frequent blood transfusions, although a few lucky patients have been cured through bone marrow transplants from perfectly matching donors.

"Platelets weave a mesh of fibrin that stems blood loss"

Blood disorders

Blood is a delicate balancing act, with the body constantly regulating oxygen flow, iron content and clotting ability. Unfortunately, there are several genetic conditions and chronic illnesses that can disturb the balance, sometimes with deadly consequences.

Sickle cell anaemia

Anaemia is the name for any blood disorder that results in a dangerously low red blood cell count. In sickle cell anaemia, which afflicts one out of every 625 children of African descent, red blood cells elongate into a sickle shape after releasing their oxygen. The sickle-shaped cells die prematurely, leading to anaemia, or sometimes lodge in blood vessels, causing terrible pain and even organ damage. Interestingly, people who carry only one gene for sickle cell anaemia are immune to malaria.

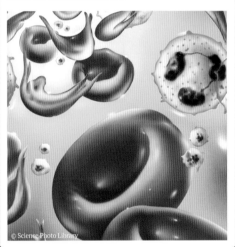
© Science Photo Library

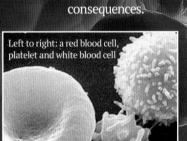
Left to right: a red blood cell, platelet and white blood cell

Hemochromatosis

One of the most common genetic blood disorders, emochromatosis is the medical term for "iron overload," in which your body absorbs and stores too much iron from food. Severity varies wildly, and many people experience few symptoms, but others suffer serious liver damage or scarring(cirrhosis), irregular heartbeat, diabetes and even heart failure. Symptoms can be aggravated by taking too much vitamin C.

Deep vein thrombosis

Thrombosis is the medical term for any blood clot that is large enough to block a blood vessel. When a blood clot forms in the large, deep veins of the upper thigh, it's called deep vein thrombosis. If such a clot breaks free, it can circulate through the bloodstream, pass through the heart and become lodged in arteries in the lung, causing a pulmonary embolism. Such a blockage can severely damage portions of the lungs, and multiple embolisms can even be fatal.

Blood and healing
More than a one-trick pony, your blood is a vital cog in the healing process

Think of blood as the body's emergency response team to an injury. Platelets emit signals that encourage blood vessels to contract, stemming blood loss. The platelets then collect around the wound, reacting with a protein in plasma to form fibrin, a tissue that weaves into a mesh. Blood flow returns and white blood cells begin their hunt for bacteria. Fibroblasts create beds of fresh collagen and capillaries to fuel skin cell growth. The scab begins to contract, pulling the growing skin cells closer together until damaged tissue is replaced.

STAGE 1
INJURY
When the skin surface is cut, torn or scraped deeply enough, blood seeps from broken blood vessels to fill the wound. To stem the flow of bleeding, the blood vessels around the wound constrict.

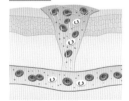

STAGE 2
HAEMOSTASIS
Activated platelets aggregate around the surface of the wound, stimulating vasoconstriction. Platelets react with a protein in plasma to form fibrin, a web-like mesh of stringy tissue.

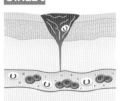

STAGE 3
INFLAMMATORY STAGE
Once the wound is capped with a drying clot, blood vessels open up again, releasing plasma and white blood cells into the damaged tissue. Macrophages digest harmful bacteria and dead cells.

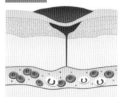

STAGE 4
PROLIFERATIVE STAGE
Fibroblasts lay fresh layers of collagen inside the wound and capillaries begin to supply blood for the forming of new skin cells. Fibrin strands and collagen pull the sides of the wound together.

Inside a blood vessel

Discover what happens every time your heart beats

Inside your body there is a vast network of blood vessels that, if laid end to end, could easily wrap twice around the Earth. They are an important part of your circulatory system, carrying the equivalent of more than 14,000 litres of blood around your body every day to transport vital nutrients to where they are needed.

There are five main types of blood vessel. In general, arteries carry oxygenated blood away from the heart and have special elastic fibres in their walls to help squeeze it along when the heart muscle relaxes. The arteries then branch off into arterioles, which pass the blood into the capillaries, tiny blood vessels that transport nutrients from the blood into the body's tissues via their very thin walls.

As well as nourishing the tissue cells, capillaries also remove their waste products, passing the now deoxygenated blood on to the venules. These vessels drain the blood into the veins, which, with the help of valves that stop the blood flowing in the reverse direction, carry it back to the heart where it can pick up more oxygen.

In contrast to the other blood vessels in the body, the pulmonary artery takes deoxygenated blood from the heart to the lungs, where it is oxygenated and carried back to the heart via the pulmonary veins.

1 Red blood cells
These disc-shaped cells contain the protein haemoglobin, which enables them to carry oxygen and carbon dioxide around your body.

3 Plasma
The liquid part of your blood is made up of water, salts and enzymes, and helps transport hormones, proteins, nutrients and waste around your body.

2 White blood cells
An important part of your immune system, some of these cells produce antibodies that defend against bacteria and viruses.

4 Platelets
These tiny cells trigger the process that causes blood to clot, helping to stop any bleeding if you are injured.

5 Vessel
Blood vessels transport blood and the nutrients it carries to the tissues around your body.

What is hyperventilation?

Discover why it's not always best to reach for the paper bag

Also known as over-breathing, hyperventilation is a common side effect of a panic attack or strong feelings of anxiety. When you feel breathless, you breathe more rapidly in an attempt to get more oxygen into your system. However, rather than increasing the levels of oxygen in your blood, this instead causes the carbon dioxide levels to decrease. As a result, the pH of your blood becomes more alkaline, causing the red blood cells to cling on to their oxygen instead of passing it on to the tissue cells as they would normally. This simply exacerbates the problem, causing you to try to breathe in more oxygen and lowering your carbon dioxide levels further.

One way to stop the vicious cycle is to breathe into a paper bag, forcing you to re-breathe some of your exhaled carbon dioxide. However, this will only work if the hyperventilation was brought on by anxiety or a panic attack. Over-breathing can also be caused by asthma, infections, bleeding or heart attacks, and in these cases, increased levels of carbon dioxide are dangerous. Therefore, the best course of treatment is to try to stay calm and slow your breathing, and seek medical help if the problem persists.

Breathing into a paper bag can be a dangerous way to treat hyperventilation

Sensitive teeth

Why can eating cold foods be painful, and what can toothpaste do to help?

If you've ever sipped on an ice-cold drink and felt a sharp pain in your teeth, then you've probably suffered from tooth sensitivity. It's a common condition caused when the soft interior layer of your teeth, known as dentin, becomes exposed, allowing hot or cold substances to reach the nerves inside.

Luckily, the problem can often be treated by simply changing your toothpaste. Special desensitising toothpastes contain ingredients that block the channels within the dentin layer, stopping anything from getting inside. However, if the problem persists, then it's best to see your dentist for alternative treatment.

Hot or cold food and drink can cause sharp pains for those with sensitive teeth

Getting to the nerves

How do teeth become sensitive?

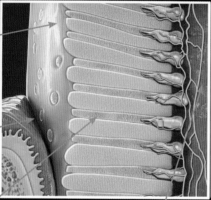
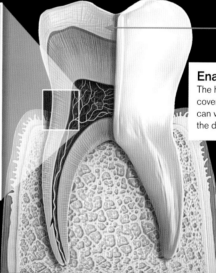

Dentin
This soft substance inside your teeth contains thousands of microscopic fluid-filled channels.

Gum tissue
The connective tissue in which your teeth grow can pull away from them, uncovering the dentin layer beneath.

Tubules
Heat or cold travels down the dentin channels towards the pulp tissue at the centre of the tooth.

Pulp tissue
Nerves inside the pulp tissue become stimulated by the external factors, causing feelings of pain.

Enamel
The hard, white shell that covers the top of your teeth can wear away, exposing the dentin layer.

Jaundice

What causes human skin to turn yellow?

Jaundice, also known as icterus, is a term used to describe the yellowing of the skin and the whites of the eyes. It is caused by a build-up of a substance called bilirubin in the blood, which is the yellow-coloured waste product produced when old red blood cells break down. Normally, bilirubin is filtered out of your bloodstream by your liver and then excreted, but people with conditions affecting their liver can be left with an excess of the substance, which can leak out into surrounding body tissue.

While adults and older children can get jaundice, it is most common in newborn babies.

As an infant's liver is still developing, it can be slower to remove bilirubin from the blood than normal, but the symptoms usually disappear after a few weeks with no need for treatment.

However, if the jaundice becomes more severe, it can be treated using phototherapy, which involves using blue coloured light to break down the bilirubin in the blood. Alternatively, the infant could receive a blood transfusion, giving them more red blood cells to counteract the bilirubin.

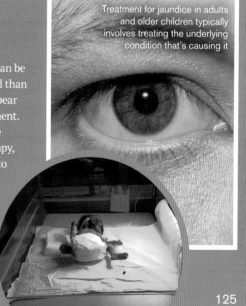

Treatment for jaundice in adults and older children typically involves treating the underlying condition that's causing it

Phototherapy can help break down the excess bilirubin in a newborn baby's blood

Hormones

How the human endocrine system develops and controls the human body

The glands in the endocrine system use chemicals called hormones to communicate with and control the cells and organs in our bodies. They are ductless glands that secrete different types of hormones directly into the bloodstream which then target specific organs.

The target organs contain hormone receptors that respond to the chemical instructions supplied by the hormone. There are 50 different types of hormone in the body and they all consist of three basic types: peptides, amines and steroids.

Steroids include the testosterone hormone. This is not only secreted by the cortex of the adrenal gland, but also from the male and female reproductive organs and by the placenta in pregnant women. The majority of hormones are called peptides that consist of short chains of amino acids. They are secreted by the pituitary and parathyroid glands. Amine hormones are secreted by the thyroid and adrenal medulla and are related to initiating the fight or flight response.

The changes that are caused by the endocrine system act more slowly than the nervous system as they regulate growth, moods, metabolism, reproductive processes and a relatively constant stable internal environment for the body (homeostasis). The pituitary, thyroid and adrenal glands then all combine to form the major elements of the body's endocrine system along with various other elements such as the male testes, the female ovaries and the pancreas.

"Amine hormones are secreted by the thyroid and adrenal medulla"

Adrenal gland

We have two adrenal glands that are positioned on top of both kidneys. The triangular-shaped glands each consist of a two-centimetre thick outer cortex that produces steroid hormones, which include testosterone, cortisol and aldosterone.

The ellipsoid shaped, inner part of the gland is known as the medulla, which produces noradrenaline and adrenaline. These hormones increase the heart rate, and the body's levels of oxygen and glucose while reducing non-essential body functions.

The adrenal gland is known as the 'fight or flight' gland as it controls how we respond to stressful situations, and prepares the body for the demands of either fighting or running away as fast as you can. Prolonged stress over-loads this gland and causes illness.

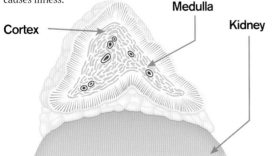

Medulla

Cortex

Kidney

The endocrine system

Thymus
Is part of the immune system. It produces thymosins that control the behaviour of white blood T-cells.

Adrenal glands
Controls the burning of protein and fat, and regulates blood pressure. The medulla secretes adrenaline to stimulate the fight or flight response.

Male testes
These two glands produce testosterone that is responsible for sperm production, muscle and bone mass and sex drive.

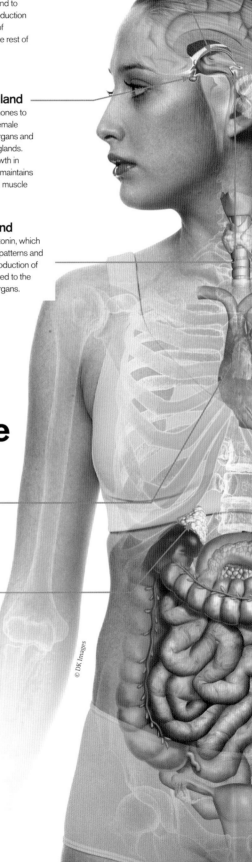

Hypothalamus
Releases hormones to the pituitary gland to promote its production and secretion of hormones to the rest of the body.

Pituitary gland
Releases hormones to the male and female reproductive organs and to the adrenal glands. Stimulates growth in childhood and maintains adult bone and muscle mass.

Pineal gland
Secretes melatonin, which controls sleep patterns and controls the production of hormones related to the reproductive organs.

© DK Images

Pituitary gland

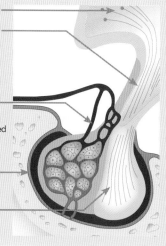

Hypothalamus

Hypothalamus neurons
These synthesise and send hormones to the posterior lobe.

Portal veins
Hormones from the hypothalamus are carried to the anterior lobe through these veins.

Anterior lobe

Posterior lobe

The pea-sized pituitary gland is a major endocrine gland that works under the control of the hypothalamus. The two organs inside an individuals brain work in concert and mediate feedback loops in the endocrine system to maintain control and stability within the body.

The pituitary gland features an anterior (front) lobe and a posterior (rear) lobe. The anterior lobe secretes growth hormones that stimulate the development of the muscles and bones; it also stimulates the development of ovarian follicles in the female ovary. In males, it is this that actually stimulates the production of sperm cells. The posterior lobe stores vasopressin and oxytocin that is supplied by the hypothalamus. Vasopressin allows the retention of water in the kidneys and suppresses the need to excrete urine. It also raises blood pressure by contracting the blood vessels in the heart and lungs.

Oxytocin influences the dilation of the cervix before giving birth and the contraction of the uterus after birth. The lactation of the mammary glands are stimulated by oxytocin when mothers begin to breastfeed.

Thyroid and parathyroids

Parathyroid
Works in combination with the thyroid to control levels of calcium.

Thyroid
Important for maintaining the metabolism of the body. It releases T3 and T4 hormones to control the breakdown of food and store it, or release it as energy.

Right lobe

Thyroid cartilage (Adam's apple)

FRONT

REAR

Left lobe

Isthmus

Trachea (windpipe)

Parathyroids

The two lobes of the thyroid sit on each side of the windpipe and are linked together by the isthmus that runs in front of the windpipe. It stimulates the amount of body oxygen and energy consumption, thereby keeping the metabolic rate of the body at the current levels to keep you healthy and active.

The hypothalamus and the anterior pituitary gland are in overall control of the thyroid and they respond to changes in the body by either suppressing or increasing thyroid stimulating hormones. Overactive thyroids cause excessive sweating, weight loss and sensitivity to heat, whereas underactive thyroids cause sensitivity to hot and cold, baldness and weight gain. The thyroid can swell during puberty and pregnancy or due to viral infections or lack of iodine in a person's diet.

The four small parathyroids regulate the calcium levels in the body; it releases hormones when calcium levels are low. If the level of calcium is too high the thyroid releases calcitonin to reduce it. Therefore, the thyroid and parathyroids work in tandem.

Pancreas
Maintains healthy blood sugar levels in the blood stream.

Pancreatic cells

Islets of Langerhans

Red blood cells

Acinar cells
These secrete digestive enzymes to the intestine.

The pancreas is positioned in the abdominal cavity above the small intestine. Consisting of two types of cell, the exocrine cells do not secrete their output into the bloodstream but the endocrine cells do.

The endocrine cells are contained in clusters called the islets of Langerhans. They number approximately 1 million cells and are only one or two per cent of the total number of cells in the pancreas. There are four types of endocrine cells in the pancreas. The beta cells secrete insulin and the alpha cells secrete glucagon, both of which stimulate the production of blood sugar (glucose) in the body. If the Beta cells die or are destroyed it causes type 1 diabetes, which is fatal unless treated with insulin injections.

The other two cells are the gamma and delta cells. The former reduces appetite and the latter reduces the absorption of food in the intestine.

Female ovaries
Are stimulated by hormones from the pituitary gland and control the menstrual cycle.

Duct cells
Secrete bicarbonate to the intestine.

Ears feed sounds to the brain but also control balance.

About 100 million photoreceptors per eye.

9,000 taste buds over the tongue and the throat.

We can process over 10,000 different smells.

Touch is the first sense to develop in the womb.

Exploring the sensory system

The complex senses of the human body and how they interact is vital to the way we live day to day

The sensory system is what enables us to experience the world. It can also warn us of danger, trigger memories and protect us from damaging stimuli, such as hot surfaces. The sensory system is highly developed, with many components detecting both physical and emotional properties of the environment. For example, it can interpret chemical molecules in the air into smells, moving molecules of sound into noises and pressure placed on the skin into touch. Indeed, some of our senses are so finely tuned that they allow reactions within milliseconds of detecting a new sensation.

The five classic senses are sight, hearing, smell, taste and touch. We need senses not only to interpret the world around us, but also to function within it. Our senses enable us to modify our movements and thoughts, and sometimes they directly feed signals into muscles. The sensory nervous system that lies behind this is made up of receptors, nerves and dedicated parts of the brain.

There are thousands of different stimuli that can trigger our senses, including light, heat, chemicals in food and pressure. These 'stimulus modalities' are then detected by specialised receptors, which convert them into sensations such as hot and cold, tastes, images and touch. The incredible receptors – like the eyes, ears, nose, tongue and skin – have adapted over time to work seamlessly together and without having to be actively 'switched on'.

However, sometimes the sensory system can go wrong. There are

hundreds of diseases of the senses, which can have both minor effects, or a life-changing impact. For example, a blocked ear can affect your balance, or a cold your ability to smell – but these things don't last for long.

In contrast, say, after a car accident severing the spinal cord, the damage can be permanent. There are some very specific problems that the sensory system can bring as well. After an amputation, the brain can still detect signals from the nerves that used to connect to the lost limb. These sensations can cause excruciating pain; this particular condition is known as phantom limb syndrome.

However the sensory system is able to adapt to change, with the loss of one often leading to others being heightened. Our senses normally function to gently inhibit each other in order to moderate individual sensations. The loss of sight from blindness is thought to lead to strengthening of signals from the ears, nose and tongue. Having said this, it's certainly not universal among the blind, being more common in people who have been blind since a young age or from birth. Similarly, some people who listen to music like to close their eyes, as they claim the loss of visual input can enhance the audio experience.

Although the human sensory system is well developed, many animals out-perform us. For example, dogs can hear much higher-pitched sounds, while sharks have a far better sense of smell – in fact, they can sniff out a single drop of blood in a million drops of water!

Body's messengers

The sensory system is formed from neurons. These are specialised nerve cells which transmit signals from one end to the other – for example, from your skin to your brain. They are excitable, meaning that when stimulated to a certain electrical/chemical threshold they will fire a signal. There are many different types, and they can interconnect to affect each other's signals.

Retinal neuron
These retinal bipolar cells are found in the eye, transmitting light signals from the rods and cones (where light is detected) to the ganglion cells, which send impulses into the brain.

Olfactory neuron
The many fine dendritic arms of the olfactory cell line the inner surface of the nasal cavity and detect thousands of different smells, or odorants.

Purkinje cell
These are the largest neurons in the brain and their many dendritic arms form multiple connections. They can both excite and inhibit movement.

Anaxonic neuron
Found within the retina of the eye, these cells lack an axon (nerve fibre) and allow rapid modification of light signals to and from bipolar cells.

Motor neuron
These fire impulses from the brain to the body's muscles, causing contraction and thus movement. They have lots of extensions (ie they are multipolar) to spread the message rapidly.

Pyramidal neuron
These neurons have a triangular cell body, and were thus named after pyramids. They help to connect motor neurons together.

Unipolar neuron
These sensory neurons transduce a physical stimulus (for example, when you are touched) into an electrical impulse.

How do we smell

Find out how our nose and brain work together to distinguish scents

Olfactory nerve
New signals are rapidly transmitted via the olfactory nerve to the brain, which collates the data with sight and taste.

Olfactory bulb
Containing many types of cell, olfactory neurons branch out of here through the cribriform plate below.

Olfactory neuron
These neurons are highly adapted to detect a wide range of different odours.

Olfactory epithelium
Lining the nasal cavity, this layer contains the long extensions of the olfactory neurons and is where chemical molecules in air trigger an electric impulse.

Cribriform plate
A bony layer of the skull with many tiny holes, which allow the fibres of the olfactory nerves to pass from nose to brain.

Total recall
Have you ever smelt something that transported you back in time? This is known as the Madeleine effect because the writer Marcel Proust once described how the scent of a madeleine cake suddenly evoked strong memories and emotions from his childhood.

The opposite type of recall is voluntary memory, where you actively try and remember a certain event. Involuntary memories are intertwined with emotion and so are often the more intense of the two. Younger children under the age of ten have stronger involuntary memory capabilities than older people, which is why these memories thrust you back to childhood. Older children use voluntary memory more often, eg when revising for exams.

A quick, sharp pain is a common triggers for a lightning reflex

Understanding lightning reflexes

Have you ever felt something scorching hot or freezing cold, and pulled your hand away without even thinking about it? This reaction is a reflex. Your reflexes are the most vital and fastest of all your senses. They are carried out by the many 'reflex arcs' located throughout the body.

For example, a temperature-detecting nerve in your finger connects to a motor nerve in your spine, which travels straight to your biceps, creating a circular arc of nerves. By only having two nerves in the circuit, the speed of the reflex is as fast as possible. A third nerve transmits the sensation to the brain, so you know what's happened, but this nerve doesn't interfere with the arc; it's for your information only. There are other reflex arcs located within your joints, so that, say, if your knee gives way or you suddenly lose balance, you can compensate quickly.

1. Touch receptor
When a touch receptor is activated, information about the stimulus is sent to the spinal cord. Reflex actions, which don't involve the brain, produce rapid reactions to dangerous stimuli.

2. Signal sent to spine
When sensory nerve endings fire, information passes through nerve fibres to the spinal cord.

3. Motor neurons feed back
The signals trigger motor neurons that initiate their own impulses that feed back to the muscle, telling it to move the body part.

Key nerves

These transmit vital sensory information to our brain while also sending motor function signals all around the body

Olfactory nerve
Starting in the nose, this nerve converts chemical molecules into electrical signals that are interpreted as distinct odours via chemoreceptors.

Optic nerve
The optic nerves convert light signals into electrical impulses, which are interpreted in the occipital lobe at the back of the brain. The resulting image is seen upside down and back to front, but the brain reorients the image.

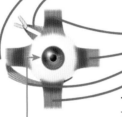

Eye movements
The trochlear, abducent and oculomotor nerves control the eye muscles and so the direction in which we look.

Trigeminal nerve
This nerve is an example of a mechanoreceptor, as it fires when your face is touched. It is split into three parts, covering the top, middle and bottom thirds of your face.

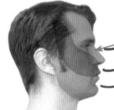

Facial and trigeminal motors
The motor parts of these nerves control the muscles of facial expression (for example, when you smile), and the muscles of the jaw to help you chew.

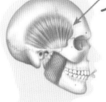

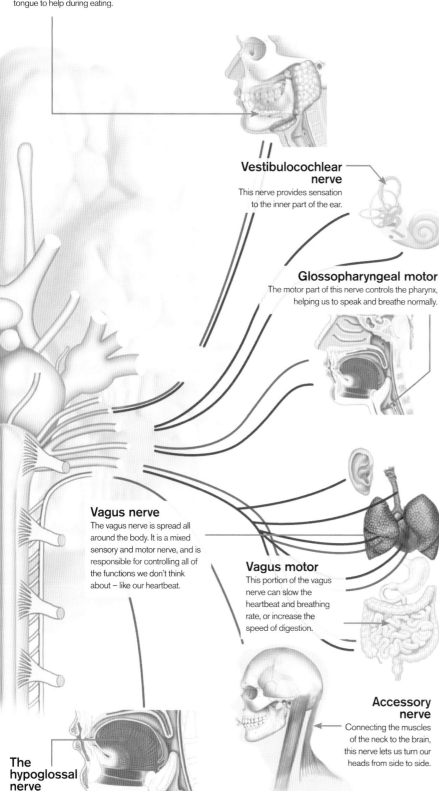

Intermediate nerve
This is a small part of the larger facial nerve. It provides the key sensation to the forward part of the tongue to help during eating.

Vestibulocochlear nerve
This nerve provides sensation to the inner part of the ear.

Glossopharyngeal motor
The motor part of this nerve controls the pharynx, helping us to speak and breathe normally.

Vagus nerve
The vagus nerve is spread all around the body. It is a mixed sensory and motor nerve, and is responsible for controlling all of the functions we don't think about – like our heartbeat.

Vagus motor
This portion of the vagus nerve can slow the heartbeat and breathing rate, or increase the speed of digestion.

Accessory nerve
Connecting the muscles of the neck to the brain, this nerve lets us turn our heads from side to side.

The hypoglossal nerve
This nerve controls the movements of the tongue.

Crossed sense

Synaesthesia is a fascinating, if yet completely understood, condition. In some people, two or more of the five senses become completely linked so when a single sensation is triggered, all the linked sensations are activated too. For example, the letter 'A' might always appear red, or seeing the number '1' might trigger the taste of apples. Sights take on smells, a conversation can take on tastes and music can feel textured.

People with synaesthesia certainly don't consider it to be a disorder or a disease. In fact, many do not think what they sense is unusual, and they couldn't imagine living without it. It often runs in families and may be more common than we think. More information about the condition is available from the UK Synaesthesia Association (www.uksynaesthesia.com).

Non-synaesthetes struggle to identify a triangle of 2s among a field of number 5s.

But a synaesthete who sees 2s as red and 5s as green can quickly pick out the triangle.

A patient's sense of proprioception is being put to the test here

Is there really a 'sixth sense'?

Our sense of balance and the position of our bodies in space are sensations we rarely think about and so are sometimes thought of as a 'sixth sense'. There is a whole science behind them though, and they are collectively called proprioception. There are nerves located throughout the musculoskeletal system (for example, within your muscles, tendons, ligaments and joints) whose job it is to send information on balance and posture back to the brain. The brain then interprets this information rapidly and sends instructions back to the muscles to allow for fine adjustments in balance. Since you don't have to think about it and you can't switch it off, you don't know how vital these systems are until they're damaged. Sadly some medical conditions, including strokes, can affect our sense of proprioception, making it difficult to stand, walk, talk and move our limbs.

CURIOUS QUESTIONS

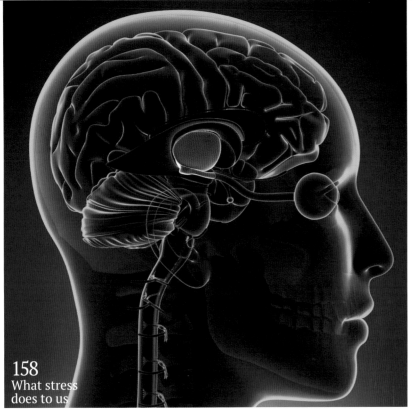

158
What stress does to us

136
What is a brain freeze?

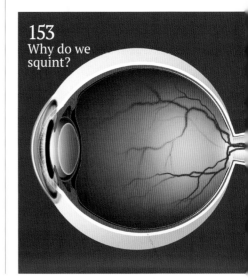

153
Why do we squint?

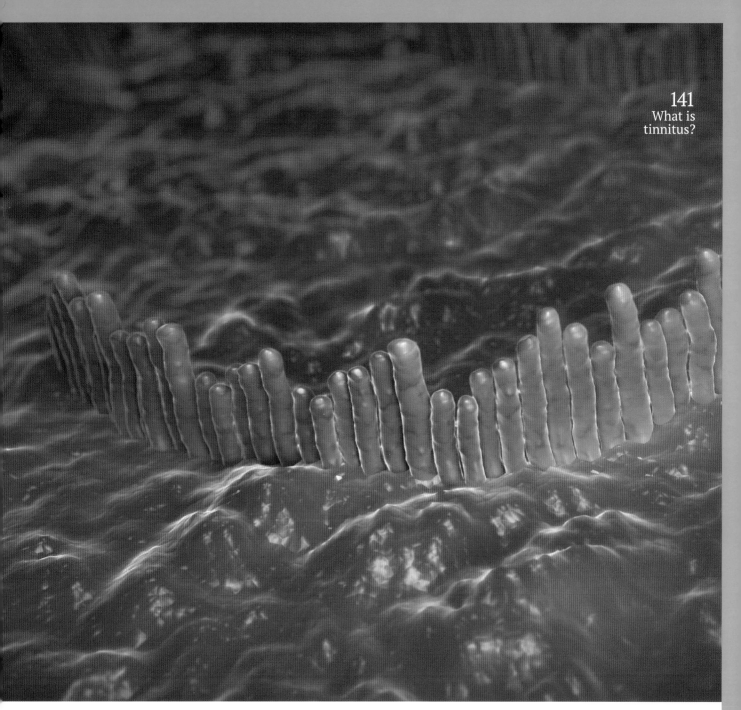

141
What is
tinnitus?

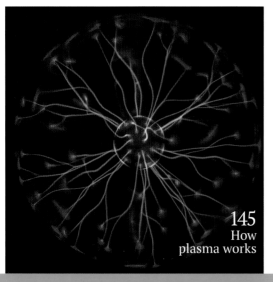

145
How
plasma works

159
What causes
insomnia?

© Thinkstock; Alamy

Left or right brained?

Actually, you're neither. Discover the truth behind the way we think

It's true that the different sides of the brain perform different tasks, but do these anatomical asymmetries really define our personalities? Some psychologists argue that creative, artistic individuals have a more developed right hemisphere, while analytical, logical people rely more heavily on the left side of the brain, but so far, the evidence for this two-sided split has been lacking.

In a study published in the journal PLOS ONE, a team at the University of Utah attempted to answer the question. They divided the brain up into 7,000 regions and analysed the fMRI scans of over 1,000 people, in order to determine whether the networks on one side of the brain were stronger than the networks on the other.

Despite the popularity of the left versus right brain myth, the team found no difference in the strength of the networks in each hemisphere, or in the amount we use either side of our brains.

Instead, they showed that the brain is more like a network of computers. Local nerves can communicate more efficiently than distant ones, so instead of sending every signal across from one hemisphere of the brain to the other, neurones that need to be in constant communication tend to develop into organised local hubs, each responsible for a different set of functions.

Hubs with related functions cluster together, preferentially developing on the same side of the brain, and allowing the nerves to communicate rapidly on a local scale. One example is language processing – in most people, the regions of the brain involved in speech, communication and verbal reasoning are all located on the left-hand side.

Some areas of the brain are less symmetrical than others, but both hemispheres are used relatively equally. There is nothing to say you can't be a brilliant scientist and a great artist.

Examining the human brain

What do the different parts of the brain actually do?

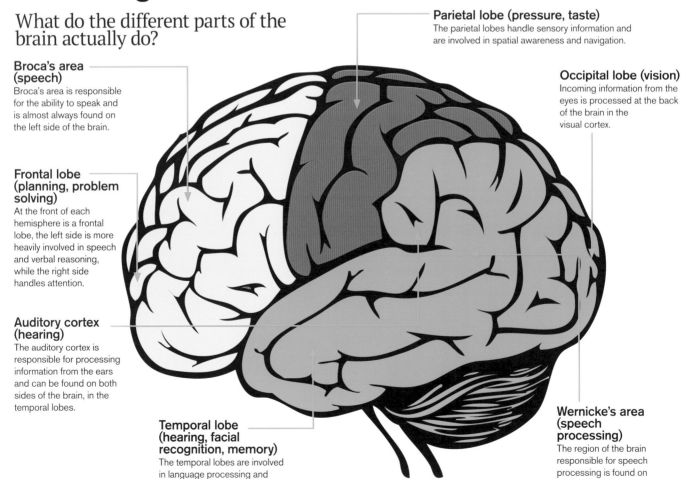

Broca's area (speech)
Broca's area is responsible for the ability to speak and is almost always found on the left side of the brain.

Frontal lobe (planning, problem solving)
At the front of each hemisphere is a frontal lobe, the left side is more heavily involved in speech and verbal reasoning, while the right side handles attention.

Auditory cortex (hearing)
The auditory cortex is responsible for processing information from the ears and can be found on both sides of the brain, in the temporal lobes.

Temporal lobe (hearing, facial recognition, memory)
The temporal lobes are involved in language processing and visual memory.

Parietal lobe (pressure, taste)
The parietal lobes handle sensory information and are involved in spatial awareness and navigation.

Occipital lobe (vision)
Incoming information from the eyes is processed at the back of the brain in the visual cortex.

Wernicke's area (speech processing)
The region of the brain responsible for speech processing is found on the left-hand side.

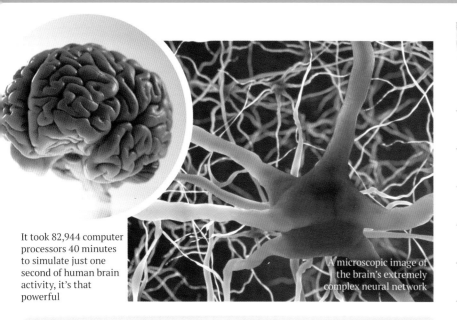

It took 82,944 computer processors 40 minutes to simulate just one second of human brain activity, it's that powerful

A microscopic image of the brain's extremely complex neural network

Myth-taken identity

The left vs right brain personality myth is actually based on Nobel Prize-winning science. In the 1940s, a radical treatment for epilepsy was trialled; doctors severed the corpus callosum of a small number of patients, effectively splitting their brains in two. If a patient was shown an object in their right field of view, they had no difficulty naming it, but if they were shown the same object from the left, they couldn't describe it. Speech and language are processed on the left side of the brain, but the information from the left eye is processed on the right. The patients were unable to say what they saw, but they could draw it. Psychologists wondered whether the differences between the two hemispheres could create two distinctive personality types, left-brained and right-brained.

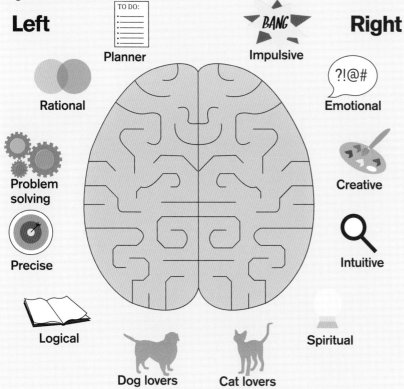

Left

Planner

Rational

Problem solving

Precise

Logical

Dog lovers

Right

BANG

Impulsive

?!@#

Emotional

Creative

Intuitive

Spiritual

Cat lovers

Give your brain a fun workout

1 Boost your memory
Look at this list of items for one minute, then cover the page and see how many you can remember:

Coin	Telephone	Grape
Duck	Potato	Pillowcase
Key	Teacup	Bicycle
Pencil	Match	Table

Difficult? Try again, but this time, make up a story in your head, linking the objects together in a narrative.

"Duck opened his front door to find his table upturned, there were teacups everywhere"

...You get the idea. Make it as silly as you like; strange things are much more memorable than the mundane.

2 Slow brain ageing
Learning a new language is one of the best ways to keep your brain active. Here are four new ways to say hello:

• Polish: Czesc!
          (che-sh-ch)
• Russian: Zdravstvuj
          (zdrah-stvooy)
• Arabic: Marhaba
          (mar-ha-ba)
• Swahili: Hujambo
          (hud-yambo)

135

What is 'brain freeze'?

That intense pain you sometimes get when you eat ice cream too fast is technically called sphenopalatine ganglioneuralgia, and it's related to migraine headaches

The pain of a brain freeze, also know as an ice cream headache, comes from your body's natural reaction to cold. When your body senses cold, it wants to conserve heat. One of the steps it takes to accomplish this is constricting the blood vessels near your skin. With less blood flowing near your skin, less heat is carried away from your core, keeping you nice and warm.

The same thing happens when something really cold hits the back of your mouth. The blood vessels in your palate constrict rapidly. When the cold goes away (because you swallowed the ice cream or cold beverage), they will rapidly dilate back to their standard, normal state.

This is harmless, but a major facial nerve called the trigeminal lies close to your palate and this nerve interprets the constriction/dilation process as pain. The location of the trigeminal nerve can cause the pain to seem like its coming from your forehead. Doctors believe this same misinterpretation of blood vessel constriction/dilation is the cause of the intense pain of a migraine headache.

A brain freeze is actually a painful side-effect of your body's survival instinct

"A major facial nerve called the trigeminal lies close to your palate"

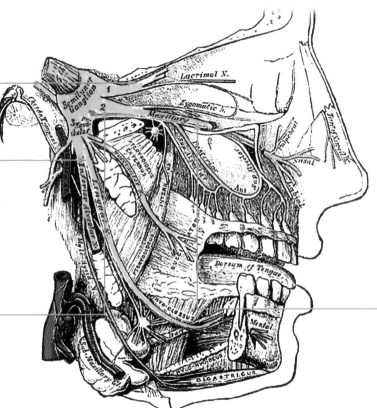

The Ophthalmic branch carries sensory messages from the eyeball, tear gland, upper nose, upper eyelid, forehead, and scalp.

The Mandibular branch carries sensory signals from the skin, teeth and gums of the lower jaw, as well as tongue, chin, lower lip and skin of the temporal region.

The trigeminal facial nerve is positioned very close to the palate. This nerve interprets palate blood vessel constriction and dilation as pain.

The Maxillary branch carries sensory messages from the skin, gums and teeth of the upper jaw, cheek, upper lip, lower nose and lower eyelid.

What makes your nose run?

Discover what is going on inside a blocked nose and why it gets runny when we're ill

I t surprises many people but the main culprit responsible for a blocked and runny nose is typically not excess mucus but swelling and inflammation.

If the nose becomes infected, or an allergic reaction is triggered, the immune system produces large quantities of chemical messengers that cause the local blood vessels in the lining of the nose to dilate. This enables more white blood cells to enter the area, helping to combat the infection, but it also causes the blood vessels to become leaky, allowing fluid to build up in the tissues.

Decongestant medicine contains a chemical that's similar to adrenaline, which causes the blood vessels to constrict, stopping them from leaking.

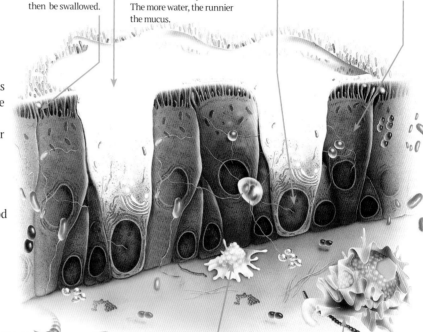

Cilia
Tiny hair-like structures move the mucus towards the back of the throat so that it can then be swallowed.

Mucus
The glycoproteins that make up mucus dissolve in water, forming a gel-like substance that traps debris. The more water, the runnier the mucus.

Goblet cell
The lining of the nose has many mucus-producing goblet cells.

Epithelial cells
The nose is lined by epithelial cells, covered in cilia.

Connective tissue
Beneath the cells lining the nose is a layer of connective tissue that is rich in blood vessels.

Macrophage
Cells of the immune system produce chemical mediators like histamine, which cause local blood vessels to become leaky.

Blood vessels
Inflammatory chemical signals cause blood vessels to dilate, allowing water to seep into the tissues, diluting the mucus and making it runny.

Bringing a patient out of a coma will not wake them up immediately

How is a person brought out of a coma?

W hen we talk about 'bringing someone out of a coma', we are referencing medically induced comas. A patient with a traumatic brain injury is deliberately put into a deep state of unconsciousness to reduce swelling and allow the brain to rest. When the brain is injured, it becomes inflamed. The swelling damages the brain because it is squashed inside the skull.

Doctors induce the coma using a controlled dose of drugs. To bring the person out of the coma, they simply stop the treatment. Bringing the patient out of the coma doesn't wake them immediately. They gradually regain consciousness over days, weeks or longer. Some people make a full recovery, others need rehabilitation or lifetime care and others may remain unaware of their surroundings.

Why do our ears 'pop' on planes?

The eardrum is a thin membrane that helps to transmit sound. Air pressure is exerted on both sides of the eardrum; with the surrounding atmospheric pressure pushing it inwards while air being delivered via a tube between the back of your nose and the eardrum pushes it outwards. This tube is called the Eustachian tube, and when you swallow it opens and a small bubble of air is able to move causing a 'pop'.

Rapid altitude changes in planes make the 'pop' much more noticeable due to bigger differences in pressure. Air pressure decreases as a plane ascends; hence air must exit the Eustachian tubes to equalise these pressures, again causing a 'pop'. Conversely, as a plane descends, the air pressure starts to increase; therefore the Eustachian tubes must open to allow through more air in order to equalise the pressure again, causing another 'pop'.

"Rapid altitude changes make the 'pop' much more noticeable"

What are freckles?

Freckles are clusters of the pigment melanin. It is produced by melanocytes deep in the skin, with greater concentrations giving rise to darker skin tones, and hence, ethnicity. Melanin protects the skin against harmful ultraviolet sunlight, but is also found in other locations around the body. Freckles are mostly genetically inherited, but not always. They become more prominent during sunlight exposure, as the melanocytes are triggered to increase production of melanin, leading to a darker complexion. People with freckles generally have pale skin tones, and if they stay in the Sun for too long they can damage their skin cells, leading to skin cancers like melanoma.

Why does hot honey and lemon help your throat when it's sore?

Honey and lemon can be drank warm as a comfort remedy, and is a popular drink with many who are feeling unwell. The idea is that honey coats the throat and therefore any inflamed areas will be 'protected' by a layer of honey, while at the same time soothing painful areas. This means it will be less painful when these areas come into contact with other surfaces when you eat or swallow. Lemon also helps to settle the stomach too, as it contains acid, which can be particularly helpful when experiencing an upset stomach from the effects of a cold or other digestion-related illness.

What is a epidural?

The science behind blocking pain explained

An epidural (meaning 'above the dura') is a form of local anaesthetic used to completely block pain while a patient remains conscious. It involves the careful insertion of a fine needle deep into an area of the spine between two vertebrae of the lower back.

This cavity is called the epidural space. Anaesthetic medication is injected into this cavity to relieve pain or numb an area of the body by reducing sensation and blocking the nerve roots that transmit signals to the brain.

The resulting anaesthetic medication causes a warm feeling and numbness leading to the area being fully anaesthetised after about 20 minutes. Depending on the length of the procedure, a top-up may be required.

This form of pain relief has been used widely for many years, particularly post-surgery and during childbirth.

1. Epidural space
The outer part of the spinal canal, this cavity is typically about 7mm (0.8in) wide in adults.

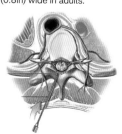

2. Epidural needle
After sterilising the area, a needle is inserted into the interspinous ligament until there is no more resistance to the injection of air or saline solution.

3. Anaesthetic
Through a fine catheter in the needle, anaesthetic is carefully introduced to the space surrounding the spinal dura.

6. Processing
Anaesthetic in the blood is filtered out by the liver and kidneys, then leaves the body in urine. The effects usually wear off a couple of hours after the initial injection.

— Liver
— Kidney
— Ureter

Bladder

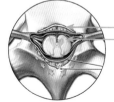

4. Absorption
Over about 20 minutes the anaesthetic medication is broken down and absorbed into the local fatty tissues.

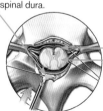

5. Radicular arteries
The anterior and posterior radicular arteries run with the ventral and dorsal nerve roots, respectively, which are blocked by the drug.

What is a memory?

Memory is the brain's ability to recall information from the past and it generally falls into three categories – sensory, short-term and long-term.

Look at this page then close your eyes and try to remember what it looks like. Your ability to recall what this page looks like is an example of your sensory memory. Depending on whether or not this page is important to you will be the determining factor in how likely it is that it will get passed on to your short-term memory.

Can you remember the last thing you did before reading this? That is your short-term memory and is a bit like a temporary storage facility where the less-important stuff can decay, whereas the more important stuff can end up in the long-term memory.

Our senses are constantly being bombarded with information. Electrical and chemical signals travel from our eyes, ears, nose, touch and taste receptors and the brain then makes sense of these signals. When we remember something, our brain refires the same neural pathways along which the original information travelled. You are almost reliving the experience by remembering it.

How does toothpaste for sensitive teeth work?

Imagine just one of your teeth. It has two primary sections: the crown located above the gum line and the root below it. The crown comprises the following layers from top to bottom: enamel, dentine and the pulp gum. Nerves branch from the root to the pulp gum. The dentine runs to the root and contains a large number of tubules or microscopic pores, which run from the outside of the tooth right to the nerve in the pulp gum.

People with sensitive teeth experience pain when their teeth are exposed to something hot, cold or when pressure is applied. Their layer of enamel may be thinner and they may have a receded gum line exposing more dentine. Therefore, the enamel and gums offer less protection and, as such, this is what makes their teeth sensitive.

Sensitive toothpaste works by either numbing tooth sensitivity, or by blocking the tubules in the dentine. Those that numb usually contain potassium nitrate, which calms the nerve of the tooth. The toothpastes that block the tubules in the dentine usually contain a chemical called strontium chloride. Repeated use builds up a strong barrier by plugging the tubules more and more.

Know your nerve cells

Take a closer look at the cells that send signals around your body

Nerve cells, or neurones, are the electrical wiring of the human body. They all have some key features in common, but depending on their specific role, they also have their own specialisms. In fact, there are more than 200 different types of neurone.

Many nerve cells can be broadly divided into four categories depending on their shape: pseudo-unipolar, bipolar, multipolar, and pyramidal. These categories are based on the number of spindly extensions that stick out from the cell body, the centre of the cell. This contains the nucleus, which carries the genetic instruction manual, and houses everything the nerve cell needs to produce the molecules that do its job. The projections link one nerve cell to the next, carrying messages in the form of electrical signals, and passing them on using chemical messengers called neurotransmitters.

There are two main types of projection. Axons are often long and tube-shaped, and carry messages away from the cell body, while dendrites are more often short and tapered, and usually receive signals from other nerve cells.

Types of neurone

The main functions of these highly specialised cells

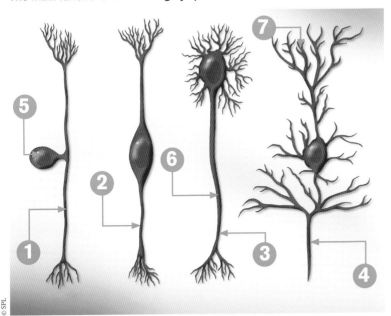

© SPL

1 Pseudo-unipolar
These cells have one projection that divides into two. The cells often transmit sensory signals.

2 Bipolar
These cells have two projections. They connect one nerve cell to the next in the brain and spinal cord.

3 Multipolar
These cells have one long projection and lots of smaller ones. They send signals to the muscles.

4 Pyramidal
These cells have lots of branching projections. They are only found in parts of the brain.

5 Cell body
The cell body is the control centre of the cell and it produces all of the proteins the cell needs.

6 Axon
There is just one axon per nerve cell. Its job is to carry electrical signals away to other cells.

7 Dendrites
Each nerve cell has hundreds or thousands of dendrites. They receive signals from other cells.

Why and how do we blush?

Blushing occurs when an excess of blood flows into the small blood vessels just under the surface of the skin. Facial skin has more capillary loops and vessels, and vessels are nearer the surface, so blushing is most visible on the cheeks, but may be seen across the whole face. The small muscles in the vessels are all controlled by the bodies nervous system.

Blushing can be affected by factors such as heat, illness, medicines, alcohol, spicy foods, allergic reactions and emotions. If you feel guilty, angry, excited or embarrassed, you will involuntarily release adrenaline, which sends the automatic nervous system into overdrive. Your breathing will increase, heart rate quicken, pupils dilate, blood will be redirected from your digestive system to your muscles, and you blush because your blood vessels dilate to improve oxygen flow around the body; this is all to prepare you for a fight or flight situation. The psychology of blushing ultimately remains elusive – some scientists even believe we have evolved to display our emotions, to act as a public apology.

"Blushing can be affected by heat, illness, medicines and spicy foods"

What makes us faint?

Fainting, or 'syncope', is a temporary loss of consciousness due to a lack of oxygen in the brain. It is preceded by dizziness, nausea, sweating and blurred vision.

The most common cause of a person fainting is overstimulation of the body's vagus nerve. Possible triggers of this include intense stress and pain, standing up for long periods or exposure to something unpleasant. Severe coughing, exercise and even urinating can sometimes produce a similar response. Overstimulation of the vagus nerve results in dilation of the body's blood vessels and a reduction of the heart rate. These two changes together mean that the body struggles to pump blood up to the brain against gravity. A lack of blood to the brain means there is not enough oxygen for it to function properly and fainting occurs.

What is tinnitus?

Find out why your ears ring after a concert

Tinnitus is a sound you can hear that isn't caused by an outside source and often manifests as a buzzing, ringing, whistling or humming noise. One of the most common causes of tinnitus is exposure to loud noises, which is why you will often experience a ringing in your ears after going to a music concert.

The loud music can temporarily damage the hair cells inside your ear and cause your brain to create phantom sounds that aren't really there. They usually disappear after a while, but prolonged exposure to loud noises can damage the hair cells permanently, resulting in a buzzing that never goes away. There is currently no cure for this type of tinnitus as the hair cells are unable to repair or replace themselves. Therefore, if you're regularly exposed to loud noises, it's important to wear earplugs to protect your delicate ears.

Loud noises are not the only cause of tinnitus, though. Other factors including a build-up of earwax, an ear infection, certain medications, a head injury or even high blood pressure, can also affect the inner workings of your ear and cause phantom sounds.

Damage to the hair cells inside your inner ear is a common cause of tinnitus

What's that buzzing?

How your ears and brain interpret real and phantom sounds

Outer ear
Sound waves enter the ear and pass through the ear canal towards the eardrum, causing it to vibrate.

Auditory nerve
The bent hairs create an electrical charge, which is carried by the auditory nerve to the brain and interpreted as sound.

Buzzing sound
When it stops receiving electrical signals, the brain spontaneously fires neurons to create phantom sounds.

Middle ear
The eardrum vibrates the ossicles (three tiny bones) to amplify the sound. The vibrations are then passed into the cochlea.

Inner ear
The vibrations cause fluid inside the cochlea to move. The fluid then rushes over and bends hair cells in the cochlea.

Cochlea damage
If the hair cells are damaged, they stop sending electrical signals to the brain.

When does your brain stop growing?

By the time a child is two years old, their brain is around 80 per cent of its adult size, but it continues to grow right up until they reach their mid-20's. However, most of this growth is not driven by the nerve cells themselves. Babies are born with almost all of the nerve cells that their brains will ever need, and the increase in size is mostly down to an increase in the number of support cells, also known as glial cells.

These fill the gaps between nerve cells, and they play a vital role in cleaning up debris, providing nutrition, and physically supporting and insulating the neurons in the brain. As children develop and get older, new connections are also made between neighbouring nerve cells, which contributes to brain growth.

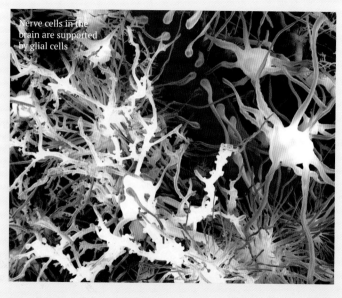
Nerve cells in the brain are supported by glial cells

What is keratin?

The secret behind some of nature's toughest materials

Keratin is a protein found in humans and animals alike. There are two main types, and each has a slightly different structure. Alpha keratin, which is the main structural component of hair, skin, nails, hooves and the wool of animals, has a coiled shape, whereas the tougher beta keratin, found in bird beaks and reptile scales, consists of parallel sheets. Both are composed of amino acids – which are the building blocks of all proteins that make up a large proportion of our cells, muscles and other tissues.

The flexibility of the keratin depends on the proportion of different amino acids present. One particular amino acid, called cysteine, is responsible for forming disulphide bridges that bond the keratin together and give it its strength. The more cysteine the keratin contains, the stronger the bonds will be, so more can be found in rigid nails and hooves than in soft, flexible hair. Incidentally, it's the sulphur within cysteine that creates the strong odour of burning hair and nails.

Curly hair has more bonds between amino acids in the protein chain that makes up keratin

Alpha keratin
How this protein makes up your hair

Alpha helix
Keratin is made of coils of amino acids held together by peptide bonds to form polypeptide chains.

Protofibril
Three alpha helices twist together to form a protofibril, the first step towards creating a hair fibre.

Microfibril
An 11-stranded cable is formed by nine protofibril joining together in a circle around two more protofibril strands.

Macrofibril
Hundreds of microfibrils bundle together in an irregular structure to create a macrofibril.

Hair cell
These macrofibrils join together within hair cells, making up the main body of the hair fibre called the cortex.

Why does hair get lighter in the summer?

Discover the secret behind why our locks lighten up in the sun

The effect of sunshine on hair is the result of ultraviolet light. The brown and red tones of skin and hair are caused by pigments known as melanin. As the short, high-energy UV wavelengths slam into the melanin pigments, they oxidise. This actually changes their chemical structure and makes them colourless.

In the skin, living cells respond to this damage by automatically producing more melanin, but there are no living cells in hair. Once the melanin is gone it cannot be replaced, and the result is gradual bleaching. Other molecules in hair can also be oxidised by UV light and as their chemical structure changes, it can make hair rough, brittle and difficult to manage.

The sun actually changes the chemical structure of the pigment in your hair

What powers your cells?

Discover how mitochondria produce all the energy your body needs

Mitochondria are known as the batteries of cells because they use food to make energy. Muscle fibres need energy for us to move and brain cells need power to communicate with the rest of the body. They generate energy, called adenosine triphosphate (ATP), by combining oxygen with food molecules like glucose.

However, mitochondria are true biological multi-taskers, as they are also involved with signalling between cells, cell growth and the cell cycle. They perform all of these functions by regulating metabolism - the processes that maintain life - by controlling Krebs Cycle which is the set of reactions that produce ATP.

Mitochondria are found in nearly every cell in your body. They are found in most eukaryotic cells, which have nucleus and other organelles bound by a cell membrane. This means cells without these features, such as red blood cells, don't contain mitochondria. Their numbers also vary based on the individual cell types, with high-energy cells, like heart cells, containing many thousands. Mitochondria are vital for most life – human beings, animals and plants all have them, although bacteria don't.

They are deeply linked with evolution of all life. It is believed mitochondria formed over a billion years ago from two different cells, where the larger cell enveloped the other. The outer cell became dependent on the inner one for energy, while the inner cell was reliant on the outer one for protection.

This inner cell evolved to become a mitochondrion, and the outer cells evolved to form building blocks for larger cell structures. This process is known as the endosymbiotic theory, which is Ancient Greek for 'living together within.'

Inside the mitochondria

Take a tour of the cell's energy factory

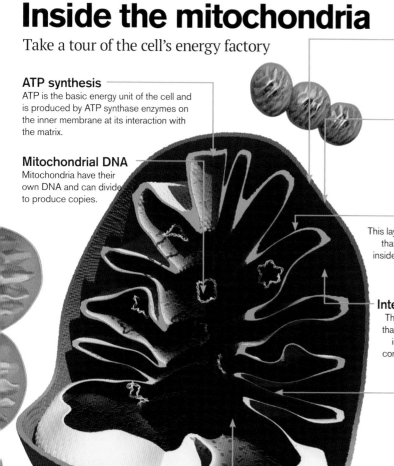

ATP synthesis
ATP is the basic energy unit of the cell and is produced by ATP synthase enzymes on the inner membrane at its interaction with the matrix.

Mitochondrial DNA
Mitochondria have their own DNA and can divide to produce copies.

Phospholipid bilayer
Every mitochondria has a double-layered surface composed of phosphates and lipids.

Outer membrane
The outer membrane contains large gateway proteins, which control passage of substances through the cell wall.

Inner membrane
This layer contains the key proteins that regulate energy production inside the mitochondria, including ATP synthase.

Inter-membrane space
This contains proteins and ions that control what is able to pass in and out of the organelle via concentration gradients and ion pumps.

Cristae
The many folds of the inner membrane increase the surface area, allowing greater energy production for high-activity cells.

Matrix
The mitochondrial matrix contains the enzymes, ribosomes and DNA, which are essential to allowing the complex energy-producing reactions to occur.

Mitochondria produce fuel for everyday activities such as exercise

How many are in a cell?

The number of mitochondria in a cell actually depends on how active that particular cell is and how much energy it requires to function. As a general rule, they can either be made up of low energy without a single mitochondrion, or made of high energy with thousands per cell. Examples of high-energy cells are heart muscles or the busy liver cells, which are still active even when you're asleep, and are packed with mitochondria to keep functioning. If you train your muscles at the gym, those cells will continue to develop mitochondria.

How anaesthesia works

By interfering with nerve transmission these special drugs stop pain signals from reaching the brain during operations

Anaesthetics are a form of drug widely used to prevent pain associated with surgery. They fall into two main categories: local and general. Local anaesthetics can be either applied directly to the skin or injected. They are used to numb small areas without affecting consciousness, so the patient will remain awake throughout a procedure.

Local anaesthetics provide a short-term blockade of nerve transmission, preventing sensory neurons from sending pain signals to the brain. Information is transmitted along nerves by the movement of sodium ions down a carefully maintained electrochemical gradient. Local anaesthetics cutoff sodium channels, preventing the ions from travelling through the membrane and stopping electrical signals travelling along the nerve.

Local anaesthesia isn't specific to pain nerves, so it will also stop information passing from the brain to the muscles, causing temporary paralysis.

General anaesthetics, meanwhile, are inhaled and injected medications that act on the central nervous system (brain and spinal cord) to induce a temporary coma, causing unconsciousness, muscle relaxation, pain relief and amnesia.

It's not known for sure how general anaesthetics 'shut down' the brain, but there are several proposed mechanisms. Many general anaesthetics dissolve in fats and are thought to interfere with the lipid membrane that surrounds nerve cells in the brain. They also disrupt neurotransmitter receptors, altering transmission of the chemical signals that let nerve cells communicate with one another.

Comfortably numb

If large areas need to be anaesthetised while the patient is still awake, local anaesthetics can be injected around bundles of nerves. By preventing transmission through a section of a large nerve, the signals from all of the smaller nerves that feed into it can't reach the brain. For example, injecting anaesthetic around the maxillary nerve will not only generate numbness in the roof of the mouth and all of the teeth on that side, but will stop nerve transmission from the nose and sinuses too. Local anaesthetics can also be injected into the epidural space in the spinal canal. This prevents nerve transmission through the spinal roots, blocking the transmission of information to the brain. The epidural procedure is often used to mollify pain during childbirth.

The body under general anaesthetic

What happens to various parts of the body when we're put under?

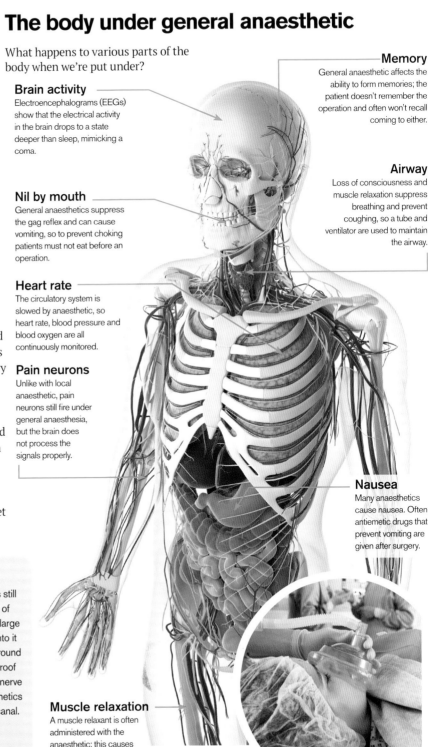

Brain activity
Electroencephalograms (EEGs) show that the electrical activity in the brain drops to a state deeper than sleep, mimicking a coma.

Nil by mouth
General anaesthetics suppress the gag reflex and can cause vomiting, so to prevent choking patients must not eat before an operation.

Heart rate
The circulatory system is slowed by anaesthetic, so heart rate, blood pressure and blood oxygen are all continuously monitored.

Pain neurons
Unlike with local anaesthetic, pain neurons still fire under general anaesthesia, but the brain does not process the signals properly.

Memory
General anaesthetic affects the ability to form memories; the patient doesn't remember the operation and often won't recall coming to either.

Airway
Loss of consciousness and muscle relaxation suppress breathing and prevent coughing, so a tube and ventilator are used to maintain the airway.

Nausea
Many anaesthetics cause nausea. Often antiemetic drugs that prevent vomiting are given after surgery.

Muscle relaxation
A muscle relaxant is often administered with the anaesthetic; this causes paralysis and enables lower doses of anaesthetic to be used.

© Getty

How decongestant medicines work

The chemicals that combat the common cold by clearing a blocked nose

W e've all had the unpleasant experience of suffering from a blocked nose that remains uncomfortably stuffy. This is one of the biggest frustrations of the common cold, but contrary to popular belief, a blocked nose is not the result of mucus. Instead, it is due to the swelling of tissues and blood vessels found in the nasal lining and sinuses, which expand and obstruct our airways.

Fortunately, decongestants can come to the rescue by providing relief from these symptoms. They contain chemicals that bind to receptors found in the nose and sinuses and cause vasoconstriction – a process where the muscles in the walls of the blood vessels contract. This reduces the size of blood vessels and so counteracts the cause of the blockage by reducing swelling.

As well as causing the contraction of blood vessels, a decongestant called pseudoephedrine is also capable of relaxing smooth muscle tissue in the airways, so you can breathe even easier.

Sinus-pressure relief
Decongestants can also be used to relieve symptoms of sinus infections.

Breathing easy
Chemicals in the decongestant help to reduce swelling in your nasal passages.

Direct delivery
Many decongestants are available as nasal sprays to provide faster relief at the source of stuffiness.

Decongestants can be found in nasal sprays as well as cold and flu relief tablets

© Thinkstock

What is plasma?

Discover the highly energised matter that powers life on Earth

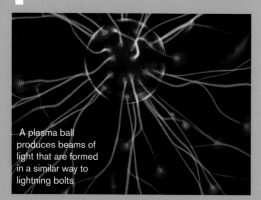

A plasma ball produces beams of light that are formed in a similar way to lightning bolts

W e're all familiar with solids, liquids and gases, which are three fundamental states of matter. But although it's not as well known, there's actually a fourth state that's more common than all of the others – plasma. This state occurs when atoms of gas are packed with energy, transforming them into separate positively and negatively charged particles. Unlike gas, plasma is a great conductor of electricity and can respond to magnetic forces. It may sound strange, but we actually see these energetic particles every day here on Earth.

During a lightning storm, for example, plasma is responsible for the beams of light we see flashing down from the sky. The massive current moving through the air energises atoms and turns them into plasma particles, which bump into each other and release light. We also see plasma every time we look at the Sun. The high temperatures are constantly converting the Sun's fuel – hydrogen and helium atoms – into positively charged ions and negatively charged electrons, making our local star the most concentrated body of plasma in the Solar System.

Why do we feel love?

The hormones and chemicals that cause us to fall head over heels

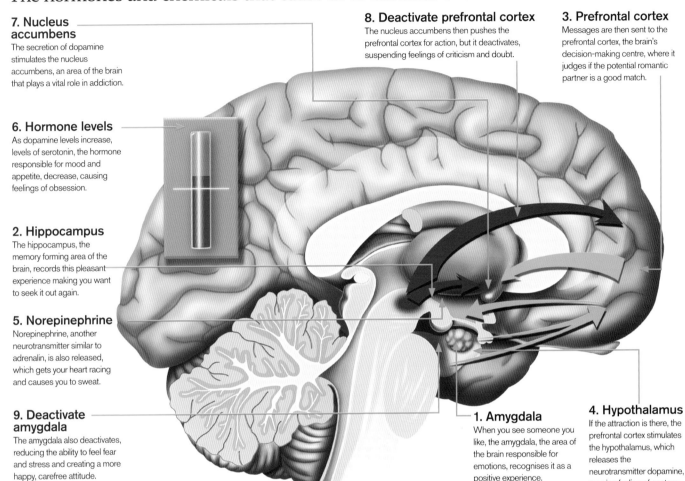

7. Nucleus accumbens
The secretion of dopamine stimulates the nucleus accumbens, an area of the brain that plays a vital role in addiction.

8. Deactivate prefrontal cortex
The nucleus accumbens then pushes the prefrontal cortex for action, but it deactivates, suspending feelings of criticism and doubt.

3. Prefrontal cortex
Messages are then sent to the prefrontal cortex, the brain's decision-making centre, where it judges if the potential romantic partner is a good match.

6. Hormone levels
As dopamine levels increase, levels of serotonin, the hormone responsible for mood and appetite, decrease, causing feelings of obsession.

2. Hippocampus
The hippocampus, the memory forming area of the brain, records this pleasant experience making you want to seek it out again.

5. Norepinephrine
Norepinephrine, another neurotransmitter similar to adrenalin, is also released, which gets your heart racing and causes you to sweat.

9. Deactivate amygdala
The amygdala also deactivates, reducing the ability to feel fear and stress and creating a more happy, carefree attitude.

1. Amygdala
When you see someone you like, the amygdala, the area of the brain responsible for emotions, recognises it as a positive experience.

4. Hypothalamus
If the attraction is there, the prefrontal cortex stimulates the hypothalamus, which releases the neurotransmitter dopamine, causing feeling of ecstasy.

How do enzymes keep you alive?

The proteins that speed up your body's chemical reactions

Enzymes such as trypsin work to help break down proteins

Enzymes increase the speed of reactions that take place inside cells by lowering the energy-activation requirement for molecular reactions. Molecules need to react with each other to reproduce, but our bodies provide neither the heat nor the pressure required for these reactions.

Each cell contains thousands of enzymes, which are amino acid strings rolled up into a ball called a globular protein. Each enzyme contains a gap called an active site into which a molecule can fit. Once inside the crack, the molecule – which becomes known as a substrate – undergoes a reaction such as dividing or merging with another molecule without having to expel energy in a collision with another molecule. The enzyme releases it and floats on within the cell's cytoplasm. The molecule and active site need to match up perfectly in order for the sped-up reaction to take place. For example, a lactose molecule would fit into a lactase enzyme's active site, but not that of a maltase enzyme.

Interestingly enough, enzymes don't actually get used up in the process, so they can then theoretically continue to be able to speed up reactions indefinitely.

© Thinkstock

Correcting heart rhythms

How can a little electricity be used to fix a heart that's beating off-kilter?

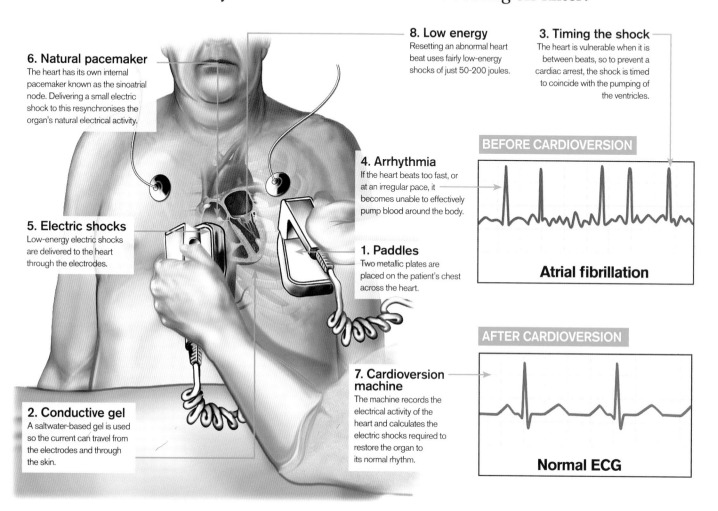

8. Low energy
Resetting an abnormal heart beat uses fairly low-energy shocks of just 50-200 joules.

3. Timing the shock
The heart is vulnerable when it is between beats, so to prevent a cardiac arrest, the shock is timed to coincide with the pumping of the ventricles.

6. Natural pacemaker
The heart has its own internal pacemaker known as the sinoatrial node. Delivering a small electric shock to this resynchronises the organ's natural electrical activity.

4. Arrhythmia
If the heart beats too fast, or at an irregular pace, it becomes unable to effectively pump blood around the body.

5. Electric shocks
Low-energy electric shocks are delivered to the heart through the electrodes.

1. Paddles
Two metallic plates are placed on the patient's chest across the heart.

2. Conductive gel
A saltwater-based gel is used so the current can travel from the electrodes and through the skin.

7. Cardioversion machine
The machine records the electrical activity of the heart and calculates the electric shocks required to restore the organ to its normal rhythm.

BEFORE CARDIOVERSION

Atrial fibrillation

AFTER CARDIOVERSION

Normal ECG

Why's salt bad for the heart?

Simply put, too much salt is bad for you as it increases the demand on your heart to pump blood around the body. This is because when you eat salt it causes retention of increased quantities of water, which increases your blood pressure, and this places more strain on your heart. Most doctors recommend moderating salt intake.

Do women have Adam's apples?

You may not realise, but actually everyone has an Adam's apple, but men's are usually easier to see in their throat. It's a bump on the neck that moves when you swallow, named after the biblical Adam. Supposedly, it's a chunk of the Garden of Eden's forbidden fruit stuck in his descendants' throats, but it's actually a bump on the thyroid cartilage surrounding the voice box. Thyroid cartilage is shield-shaped and the Adam's apple is the bit at the front.

Why do men's Adam's apples stick out more? This is partly because they have bonier necks, but it is also because their larynxes grow differently from women's during puberty to accommodate their longer, thicker vocal cords, which give them deeper voices.

What causes a rumbling stomach?

Discover how the small intestine is really to blame...

Waves of involuntary muscle contractions called peristalsis churn the food we eat to soften it and transport it through the digestive system. The contractions are caused by strong muscles in the oesophagus wall, which take just ten seconds to push food down to the stomach. Muscles in the stomach churn food and gastric juices to break it down further.

Then, after four hours, the semi-digested liquefied food moves on to the small intestine where yet more powerful muscle contractions force the food down through the intestine's bends and folds. This is where the rumbling occurs. Air from gaseous foods or that swallowed when we eat – often due to talking or inhaling through the nose while chewing food – also ends up in the small intestine, and it's this combination of liquid and gas in a small space that causes the gurgling noise.

Rumbling is louder the less food present in the small intestine, which is partly why people associate rumbling tummies with hunger. The other reason is that although the stomach may be clear, the brain still triggers peristalsis at regular intervals to rid the intestines of any remaining food. This creates a hollow feeling that causes you to feel hungry.

Oesophagus
This muscular pipe connects the throat to the stomach.

Stomach
Food is churned and mixed with gastric juices to help it to break down.

Large intestine
Food passes from the small intestine to the large intestine where it is turned into faeces.

Small intestine
Here, liquid food combined with trapped gases can make for some embarrassing noises.

Are seasickness and altitude sickness the same thing?

No, they're not – altitude sickness is a collection of symptoms brought on when you're suddenly exposed to a high-altitude environment with lower air pressure, so less oxygen enters our body. The symptoms can include a headache, fatigue, dizziness and nausea.

Seasickness, on the other hand, is a more general feeling of nausea that's thought to be caused when your brain and senses get 'mixed signals' about a moving environment – for instance, when your eyes tell you that your immediate surroundings (such as a ship's cabin) are still as a rock, while your sense of balance (and your stomach!) tells you something quite different.

This is the reason why closing your eyes or taking a turn out on deck will often help, as it reconciles the two opposing sensations.

© Thinkstock

What are blisters?

Why do burns cause bubbles to develop below the surface of the skin?

Though our skin is an amazing protector against the elements, it can become damaged by such factors as heat, cold, friction, chemicals, light, electricity and radiation, all of which 'burn' the skin. A blister is the resulting injury that develops in the upper layers of the skin.

The most common example of a blister, which we've no doubt all experienced at some time, is due to the repeated friction caused by the material of a pair of shoes rubbing against, and irritating, the skin. The resulting water blister is a kind of plasma-filled bubble that appears just below the top layers of your skin. The plasma, or serum – which is a component of your blood – is released by the damaged tissue cells and fills the spaces between the layers of skin to cushion the underlying skin and protect it from further damage. As more and more serum pours into the space, the skin begins to inflate under the pressure, forming a small balloon full of the serous liquid. Given time to heal, the skin will reabsorb the plasma after about 24 hours.

Similarly, a blood blister is a variation of the same injury where the skin has been forcefully pinched or crushed but not pierced, causing small blood vessels to rupture, leaking blood into the skin. All blisters can be tender but should never be popped to drain the fluid as this leaves the underlying skin unprotected and invites infection into the open wound.

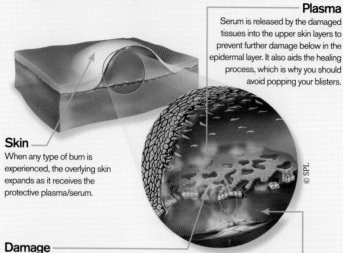

Blister caused by second-degree burns

Plasma
Serum is released by the damaged tissues into the upper skin layers to prevent further damage below in the epidermal layer. It also aids the healing process, which is why you should avoid popping your blisters.

© SPL

Skin
When any type of burn is experienced, the overlying skin expands as it receives the protective plasma/serum.

Damage
This particular example of a blister burn has caused damage to the keratinocytes in the skin. Second-degree burns are most often caused when the skin comes into contact with a hot surface, such as an iron or boiling water, or even after exposure to excessive sunlight.

Fluid reabsorbed
After a day or so the serum will be absorbed back into the body and the raised skin layers will dry out and flake off in their own time.

How a bruise forms

The colour-changing contusions caused by knocks and bumps

Whether it's a nasty fall or an accidental encounter with the edge of a table, the evidence of your mishaps can often stay with you for weeks in the form of a bruise. These contusions of the skin are caused by blood vessels bursting beneath the surface, resulting in a colourful mark that is tender to the touch.

To minimise bruising after an injury, it is best to put an ice pack on the affected area. The cold will reduce blood flow to that area, limiting the amount that can leak from the blood vessels.

Luckily our bodies are pretty good at repairing themselves and as a bruise starts to heal, it puts on an impressive colour display. After two to three weeks of changing from red to blue, then green, yellow and finally brown, it will disappear completely.

However, if a bruise doesn't fade, then your body may have blocked off a pool of blood beneath the skin, forming what is known as a haematoma.

Underneath the surface

How a blow to the skin can leave you bruised

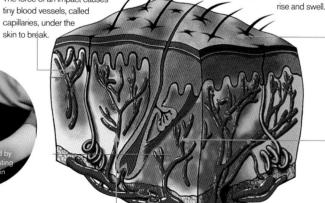

Swelling
Sometimes the blood can pool underneath your skin, causing it to rise and swell.

Burst blood vessels
The force of an impact causes tiny blood vessels, called capillaries, under the skin to break.

A bruise is caused by blood vessels bursting beneath your skin

Fading bruise
Gradually your body breaks down and reabsorbs the blood, causing the bruise to disappear.

Leaking blood
The blood inside the capillaries leaks into the soft tissue under your skin, causing it to become discoloured.

Do we control our brains or do our brains control us?

An experiment at the Max Planck Institute, Berlin, in 2008 showed that when you decide to move your hand, the decision can be seen in your brain, with an MRI scanner, before you are aware you have made a decision. The delay is around six seconds. During that time, your mind is made up but your consciousness doesn't acknowledge the decision until your hand moves. One interpretation of this is that your consciousness – the thing you think of as 'you' – is just a passenger inside a deterministic automaton. Your unconscious brain and your body get on with running your life, and only report back to your conscious mind to preserve a sense of free will. But it's just as valid to say that when you make a decision, there's always background processing going on, which the conscious mind ignores for convenience. In the same way, your eye projects an upside-down image onto your retina, but your unconscious brain turns it the right way around.

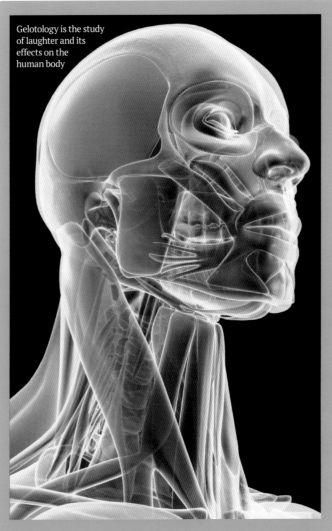

Gelotology is the study of laughter and its effects on the human body

© Alamy

What happens when we laugh?

Which muscles react when we find something funny and why is laughter so hard to fake?

Laughing can sometimes be completely involuntary and involves a complex series of muscles, which is why it's so difficult to fake and also why an active effort is required to suppress laughter in moments of sudden hilarity at inopportune moments.

In the face, the zygomaticus major and minor anchor at the cheekbones and stretch down towards the jaw to pull the facial expression upward; on top of this, the zygomaticus major also pulls the upper lip upward and outward.

The sound of our laugh is produced by the same mechanisms which are used for coughing and speaking: namely, the lungs and the larynx. When we're breathing normally, air from the lungs passes freely through the completely open vocal cords in the larynx. When they close, air cannot pass, however when they're partially open, they generate some form of sound. Laughter is the result when we exhale while the vocal cords close, with the respiratory muscles periodically activating to produce the characteristic rhythmic sound of laughing.

The risorius muscle is used to smile, but affects a smaller portion of the face and is easier to control than the zygomatic muscles. As a result, the risorius is more often used to feign amusement, hence why fake laughter is easy to detect by other humans.

What is the maximum distance the human eye can see?

Dust, water vapour and pollution in the air will rarely let you see more than 20 kilometres, even on a clear day. Often, the curvature of the Earth gets in the way first – eg at sea level, the horizon is only 4.8km away. On the top of Mt Everest, you could theoretically see for 339km, but in practice cloud gets in the way. For a truly unobstructed view, look up. On a clear night, you can see the Andromeda galaxy with the naked eye, which is 2.25 million light years away.

Our line of sight can be impeded by many things, from pollution to the curvature of the Earth

What is dandruff?

Dandruff is when dead skin cells fall off the scalp. This is normal, as our skin is always being renewed. About half the population of the world suffers from an excessive amount of this shedding, which can be triggered by things like temperature or the increased activity of a microorganism that normally lives in everyone's skin, known as malassezia globosa. Dandruff is not contagious and there are many treatments available, the most common is specialised shampoo.

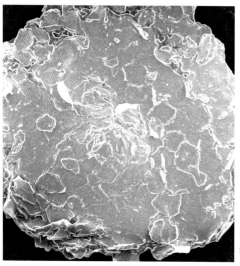

© Horoporo

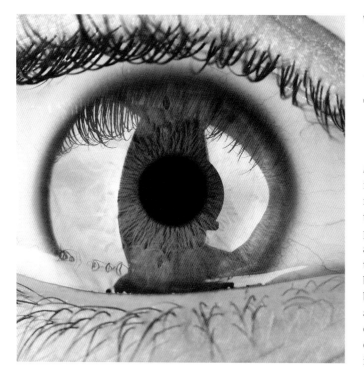

Why do eyes take a while to adjust to dark?

At the back of the eye on the retina, there are two types of photoreceptors (cells which detect light). Cones deal with colour and fine detail and act in bright light, while rods deal with vision in low-light situations. In the first few minutes of moving into a dark room, cones are responsible for vision but provide a poor picture. Once the rods become more active, they take over and create a much better picture in poor light. Once you move back into light, the rods are reset and so dark-adaption will take a few moments again. Soldiers are trained to close or cover one eye at night when moving in and out of a bright room, or when using a torch, to protect their night vision. Once back in the dark, they reopen the closed eye with the rods still working and, as a result, maintain good vision. This allows them to keep operating in a potentially hostile environment at peak operational efficiency. Give it a try next time you get up in the middle of the night, it may help you avoid tripping over in the dark.

Allergies can be a real nuisance in people's lives, but they can be controlled

Why do some people have allergies and some don't?

Allergies can be caused by two things: host and environmental factors. Host is if you inherit an allergy from your parents or are likely to get it due to your age, sex or racial group. Environmental factors can include things such as pollution, epidemic diseases and your diet.

People who are most likely to develop allergies have a condition known as 'atopy'. Atopy is not an illness but an inherited feature, which makes individuals more likely to develop an allergic disorder. Atopy tends to run in families.

The reason why atopic people have a tendency to develop allergic disorders is because they have the ability to produce the allergy antibody called 'Immunoglobulin E' or 'IgE' when they come into contact with a particular substance. However, not everyone who has inherited the tendency to be atopic will develop an allergic disorder.

"People who are likely to develop allergies have a condition known as 'atopy'"

Eczema explained

What causes the skin to react to otherwise harmless material?

Eczema is a broad term for a range of skin conditions, but the most common form is atopic dermatitis. People with this condition have very reactive skin, which mounts an inflammatory response when in contact with irritants and allergens. Mast cells release histamine, which can lead to itching and scratching, forming sores open to infection.

There is thought to be a genetic element to the disease and a gene involved in retaining water in the skin has been identified as a potential contributor, but there are many factors.

Eczema can be treated with steroids, which suppress immune system activity, dampening the inflammation so skin can heal. In serious cases, immunosuppressant drugs – used to prevent transplant rejection – can actually be used to weaken the immune system so it no longer causes inflammation in the skin.

The histamine increase can cause itching, leading to open sores

Under the skin

What happens inside the body when eczema flares up?

Ceramides
The membranes of skin cells contain waxy lipids to prevent moisture evaporation, but these are often deficient in eczema.

Allergen
Eczema is commonly triggered by the same things as many allergies – anything from pet hair to certain types of food.

Allergen entry route
The cells of the skin are normally tightly bound together to prevent contaminants from entering the body, but in eczema there are gaps.

Water loss
The skin is less able to retain water, leading to dryness and irritation.

Inflammatory response
The immune system produces a response to allergens beneath the skin, leading to redness, itching and also inflammation.

© Alamy; J Kadavoor; Thinkstock

What are growing pains?

The medical name for growing pains is 'recurrent nocturnal limb pain in children', and it describes the sensation of aching, crampy pain most often felt at night in the lower half of the legs.

Children and preteens are often told that they experience these aches and pains because they are growing, but this is untrue. If the pain really were caused by growth itself, doctors would expect to be visited by children that were going through a growth spurt, but there does not seem to be any link between periods of rapid bone growth and experience of 'growing pains'.

The pain is not in the bones or joints but is actually in the muscles and soft tissues, and one of the best explanations is that the pain is the result of strain or overuse of the muscles and joints during the day.

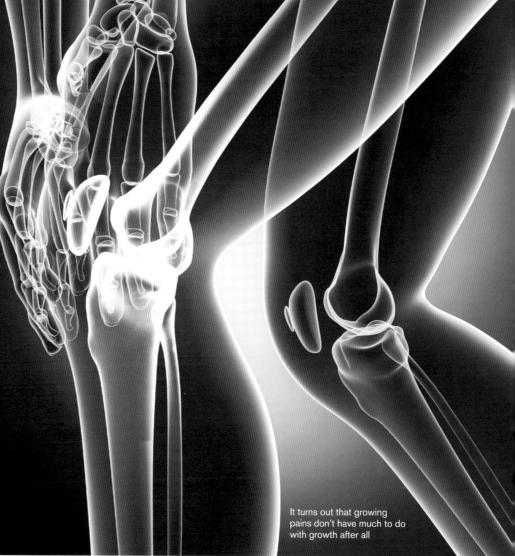

It turns out that growing pains don't have much to do with growth after all

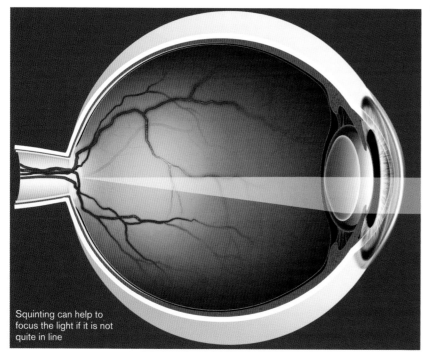

Squinting can help to focus the light if it is not quite in line

Why can we see clearer when we squint?

It doesn't work for everyone, but for some people things come into focus when they half close their eyes. This is because of the way that the eye focuses light.

A flexible lens bends the light as it passes into the eye, focusing it on a highly sensitive spot on the retina, called the fovea. The lens changes shape depending on the distance to the object, ensuring that the light is always concentrated on this spot.

As we get older, the lens becomes less flexible and cannot focus the light as well. By half closing our eyelids, we can put a little pressure on our eyeballs, changing their shape manually and helping to bring the light into focus.

How do alveoli help you breathe?

The lungs are filled with tiny balloon-like sacs that keep you alive

Gas exchange occurs in the lungs, where toxic gases (carbon dioxide) are exchanged for fresh air with its unused oxygen content. Of all the processes in the body that keep us functioning and alive, this is the most important. Without it, we would quickly become unconscious through accumulation of carbon dioxide within the bloodstream, which would poison the brain.

The two lungs (left and right) are made up of several lobes, and the fundamental building blocks of each are the tiny alveolus. They are the final point of the respiratory tract, as the bronchi break down into smaller and smaller tubes, leading to the alveoli, which are grouped together and look like microscopic bunches of grapes. Around the alveoli is the epithelial layer – which is amazingly only a single cell thick – and this is surrounded by extremely small blood vessels called capillaries. It is here that vital gas exchange takes place between the fresh air in the lungs and the deoxygenated blood within the capillary venous system on the other side of the epithelial layer.

The alveoli of the lungs have evolved to become specialised structures, maximising their efficiency. Their walls are extremely thin and yet very sturdy. Pulmonary surfactant is a thin liquid layer made from lipids and proteins that coats of all the alveoli, reduces their surface tension and prevents them crumpling when we breathe out. Without them, the lungs would collapse.

Alveoli anatomy

How alveoli enable gas exchange

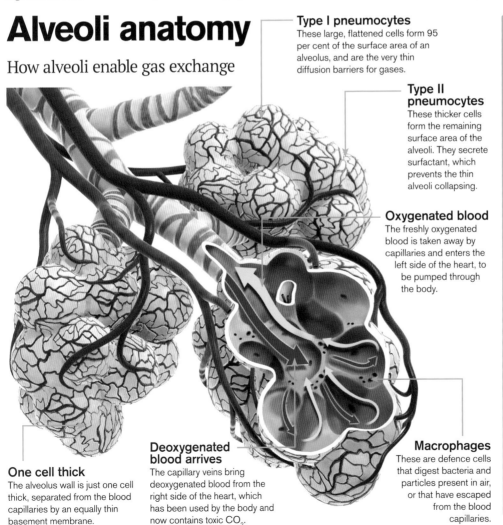

Type I pneumocytes
These large, flattened cells form 95 per cent of the surface area of an alveolus, and are the very thin diffusion barriers for gases.

Type II pneumocytes
These thicker cells form the remaining surface area of the alveoli. They secrete surfactant, which prevents the thin alveoli collapsing.

Oxygenated blood
The freshly oxygenated blood is taken away by capillaries and enters the left side of the heart, to be pumped through the body.

One cell thick
The alveolus wall is just one cell thick, separated from the blood capillaries by an equally thin basement membrane.

Deoxygenated blood arrives
The capillary veins bring deoxygenated blood from the right side of the heart, which has been used by the body and now contains toxic CO_2.

Macrophages
These are defence cells that digest bacteria and particles present in air, or that have escaped from the blood capillaries.

Breathe in, breathe out

The alveoli function to allow gas exchange, but since they're so small, they can't move new air inside and out from the body without help. That's what your respiratory muscles and ribs do, hence why your chest moves as you breathe. The diaphragm, which sits below your heart and lungs but above your abdominal organs, is the main muscle of respiration. When it contracts, the normally dome-shaped diaphragm flattens and the space within the chest cavity expands. This reduces the pressure compared to the outside atmosphere, so air rushes in. When the diaphragm relaxes, it returns to its dome shape, the pressure within the chest increases and the old air – now full of expired carbon dioxide – is forced out again. The muscles between the ribs (called intercostal muscles) are used when forceful respiration is required, such as during exercise. Try taking a deep breath and observe how both your chest expands to reduce the pressure!

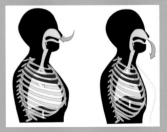

How do dilating eye drops work?

Discover how they are used to diagnose and treat eye conditions

Sight is one our most important senses, so maintaining good eye health is absolutely essential. However, eyesight problems can be difficult to detect or treat on the surface, so specialist eye doctors will often use dilating eye drops in order to get a better look inside the eye at the lens, retina and optic nerve.

The drops contain the active ingredient atropine, which works by temporarily relaxing the muscle that constricts the pupil, enabling it to remain enlarged for a longer period of time so a thorough examination can be performed. Some dilating eye drops also relax the muscle that focuses the lens inside the eye, which allows an eye doctor or optometrist to measure a prescription for young children who can't perform traditional reading tests.

Dilating eye drops are not only used to help perform procedures, they may also be administered after treatment, as they can prevent scar tissue from forming. They are also occasionally prescribed to children with lazy-eye conditions, as they will temporarily blur vision in the strong eye, causing the brain to use and strengthen the weaker eye.

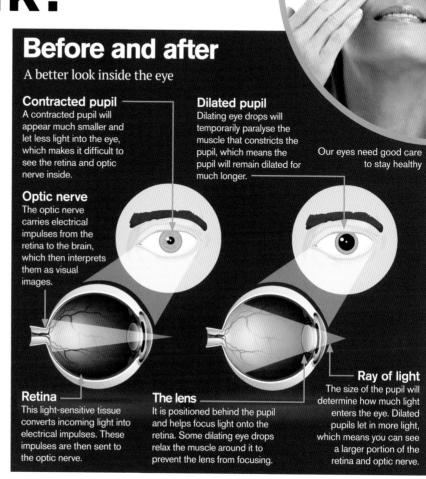

Before and after
A better look inside the eye

Contracted pupil
A contracted pupil will appear much smaller and let less light into the eye, which makes it difficult to see the retina and optic nerve inside.

Dilated pupil
Dilating eye drops will temporarily paralyse the muscle that constricts the pupil, which means the pupil will remain dilated for much longer.

Our eyes need good care to stay healthy

Optic nerve
The optic nerve carries electrical impulses from the retina to the brain, which then interprets them as visual images.

Retina
This light-sensitive tissue converts incoming light into electrical impulses. These impulses are then sent to the optic nerve.

The lens
It is positioned behind the pupil and helps focus light onto the retina. Some dilating eye drops relax the muscle around it to prevent the lens from focusing.

Ray of light
The size of the pupil will determine how much light enters the eye. Dilated pupils let in more light, which means you can see a larger portion of the retina and optic nerve.

Why do we get migraines?

Discover how these mega-headaches strike

Those who suffer from migraines know they are a constant concern as they are liable to strike at any time. Essentially, a migraine is an intense pain at the front or on one side of the head. This usually takes the form of a heavy throbbing sensation and can last as little as an hour or two and up to a few days. Other symptoms of a migraine include increased sensitivity to light, sound and smell, so isolation in a dark and quiet room often brings relief. Nausea and vomiting is also often reported, with pain sometimes subsiding after the sufferer has been sick (vomited).

It is thought that migraines occur when levels of serotonin in the brain drop rapidly. This causes blood vessels in the cortex to narrow, which is caused by the brain spasming. The blood vessels will then widen again in response, causing the intense headache. Emotional upheaval is often cited as a cause for the drop in serotonin in the brain, as is a diet in which blood-sugar levels rise and fall dramatically. Keeping stress levels low and eating healthily can help.

What are 'pins and needles'?

The numb sensation of your leg 'going to sleep' isn't caused by cutting off the blood circulation. It's actually the pressure on the nerves that is responsible. This squeezes the insulating sheath around the nerve and 'shorts it out', blocking nerve transmission. When pressure is released, the nerves downstream from the pinch point suddenly all begin firing at once. This jumble of unco-ordinated signals is a mixture of pain and touch, hot and cold all mixed together, which is why it's excruciating.

"This squeezes the insulating sheath around the nerve and 'shorts it out'"

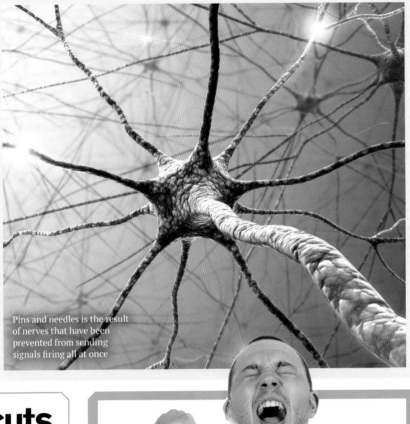

Pins and needles is the result of nerves that have been prevented from sending signals firing all at once

© Thinkstock

Why do paper cuts hurt so much?

Paper can cut your skin as it is incredibly thin and, if you were to look at it under a high-powered microscope, it has serrated edges. Critically though, a sheet of loose paper is far too soft and flexible to exert enough pressure to pierce the skin, hence why they are not a more frequent occurrence. However, if the paper is fixed in place – maybe by being sandwiched within a pack of paper – a sheet can become stiff enough to attain skin-cutting pressure. Paper

cuts are so painful once inflicted as they stimulate a large number of pain receptors – nociceptors send nerve signals to the spinal cord and brain – in a very small area due to the razor-type incision. Because paper cuts tend not to be deep, bleeding is limited, leaving pain receptors open to the environment.

Are there other 'funny bones' in the body?

The term 'funny bone' is misleading because it refers to the painful sensation you experience when you trap your ulnar nerve between the skin and the bones of the elbow joint. This happens in the so-called cubital tunnel, which directs the nerve over the elbow but has little padding to protect against external impacts. The ulnar nerve takes its name from the ulna bone, which is one of two bones that runs from the wrist to the elbow; the other is the radial bone, or radius.

No other joint in the human skeleton combines these conditions and duplicates the this erroneously named reaction so we only have one 'funny bone'.

Why do our muscles ache?

Learn what causes stiffness and pain in our muscles for days after exercise

Normally, when our muscles contract they shorten and bulge, much like a bodybuilder's biceps. However, if the muscle happens to be stretched as it contracts it can cause microscopic damage.

The quadriceps muscle group located on the front of the thigh is involved in extending the knee joint, and usually contracts and shortens to straighten the leg. However, when walking down a steep slope, say, the quadriceps contract to support your body weight as you step forward, but as the knee bends, the muscles are pulled in the opposite direction. This tension results in tiny tears in the muscle and this is the reason that downhill running causes so much delayed-onset muscle pain.

At the microscopic level, a muscle is made up of billions of stacked sarcomeres, containing molecular ratchets that pull against one another to generate mechanical force. If the muscle is taut as it tries to contract, the sarcomeres get pulled out of line, causing microscopic damage. The muscle becomes inflamed and fills with fluid, causing stiffness and activating pain receptors – hence that achy feeling you get after unfamiliar exercise.

Weight lifting and the body

What happens to your biceps when you pump iron?

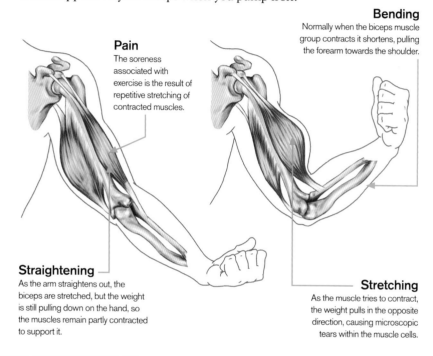

Pain
The soreness associated with exercise is the result of repetitive stretching of contracted muscles.

Bending
Normally when the biceps muscle group contracts it shortens, pulling the forearm towards the shoulder.

Straightening
As the arm straightens out, the biceps are stretched, but the weight is still pulling down on the hand, so the muscles remain partly contracted to support it.

Stretching
As the muscle tries to contract, the weight pulls in the opposite direction, causing microscopic tears within the muscle cells.

The fat hormone

Discover how the body manages to keep track of its energy reserves

The leptin (LEP) gene was originally discovered when a random mutation occurred in mice, making them put on weight

In order to know how much food to eat, the human body needs a way of assessing how much energy it currently has in storage. Leptin – more commonly known as the 'fat hormone' – essentially acts as our internal fuel gauge. It is made by fat cells and tells the brain how much fat the body contains, and whether the supplies are increasing or being used up.

Food intake is regulated by a small region of the brain called the hypothalamus. When fat stores run low and leptin levels drop, the hypothalamus stimulates appetite in an attempt to increase food intake and regain lost energy. When leptin levels are high, appetite is suppressed, reducing food intake and encouraging the body to burn up fuel.

It was originally thought that leptin could be used as a treatment for obesity. However, although it is an important regulator of food intake, our appetite is affected by many other factors, from how full the stomach is to an individual's emotional state or their food preferences. For this reason, it's possible to override the leptin message and gain weight even when fat stores are sufficient.

© Alamy

157

Why do the upper arm and upper leg have only one bone?

The makeup of the human skeleton is a fantastic display of evolution that has left us with the ability to perform incredibly complex tasks without even thinking about them. There are several different types of joint between bones in your body, which reflect their function; some are strong and allow little movement, others are weak but allow free movement. The forearm and lower leg have two bones, which form plane joints at the wrist and ankle. This allows for a range of fine movements, including gliding and rotation. The hinge joints at your elbows and knees allow for less lateral movement, but they are strong. Shoulders and hips, are ball-and-socket joints, allowing for a wide range of motion.

Why do my knuckles crack more when it's cold?

In 2015, researchers at the University of Alberta, Canada showed once and for all that the cracking sound made in finger joints is down to the formation of bubbles. As you pull, the surfaces of the joint come apart and a cavity appears in the fluid between. This makes the noise. To crack your knuckles again, you have to wait for the bubble to disappear. The researchers didn't look at the effect of climate, but it could be that something about the cold effects the behaviour of the fluid in your joints, which helps the bubbles to disperse even more rapidly.

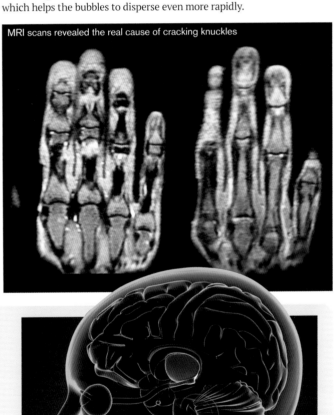

MRI scans revealed the real cause of cracking knuckles

How does stress affect the body?

The hypothalamus is a small structure that sits in the middle of the brain. It makes two key chemicals that kick-start the stress response: corticotropin-releasing hormone and vasopressin. Corticotropin-releasing hormone, as the name suggests, triggers the release of a second chemical called corticotropin. This travels in the bloodstream to the adrenal glands, which sit on top of the kidneys, and signals for them to make the steroid hormone cortisol.

Cortisol is also known as the 'stress hormone', and it has effects all across the body. It helps to return systems to normal during times of stress, including raising blood sugar, balancing pH and suppressing the immune system. Vasopressin also travels in the blood to the kidneys, but its function is slightly different. It increases the re-uptake of water, decreasing the amount of urine produced and helping the body to hold on to the reserves that it has.

The hypothalamus is the control centre of the stress response in the brain

What causes insomnia?

Why checking your phone before bed could be spoiling your sleep

" All forms of light, both natural and artificial, affect our body clock "

Most of us experience insomnia at some point in our lives, finding it difficult to drift off and stay asleep, despite having plenty of opportunity to. Typical causes of insomnia include stress and anxiety, but did you know that your gadgets could be to blame, too?

Our sleepiness and wakefulness throughout the day and night is regulated by our circadian rhythm. This is essentially our body clock, creating physical, mental and behavioural changes that occur in our bodies over a roughly 24-hour cycle. Circadian rhythms are found in most living things, including animals, plants and many tiny microbes, and they are created by natural factors in the body. However, they also respond to signals from the environment, such as light, so that we remain in sync with the Earth's rotation.

All forms of light, both natural and artificial, affect our body clock, as when the photosensitive retinal ganglion cells in our eyes detect light, they send this information to the suprachiasmatic nucleus (SCN) . When light is detected, the SCN will delay the production of melatonin, a hormone that sends us to sleep. However, the retinal ganglion cells have been found to be particularly sensitive to the blue light with a short wavelength of 480 nanometres emitted by most computer, smartphone and tablet screens. Exposure to a lot of this type of light in the hours before we go to bed has been proven to suppress melatonin levels, making it difficult for us to get to sleep.

Light sensitivity
How light affects your ability to sleep

Suprachiasmatic nucleus
The suprachiasmatic nucleus is a tiny area of neurons, located in the hypothalamus area of the brain, which controls circadian rhythms.

Light sensitivity

Pineal gland

The ganglion layer
The retina of the eye contains a layer of photosensitive ganglion cells, which contain a photopigment melanopsin, called the ganglion layer.

Optic nerve
The photosensitive ganglion cells have long fibres that connect to the optic nerve and eventually reach the suprachiasmatic nucleus.

Melatonin
When the photosensitive ganglion cells detect darkness, a message is sent to the pineal gland to produce melatonin, a hormone that can cause drowsiness.

Blocking blue light

The best way to reduce your exposure to blue light is to avoid staring at a screen within two hours prior to going to bed. Instead, illuminate the room with the warmer, longer-wavelength light from regular incandescent bulbs or even candles. However, if you just can't resist staring at your computer or phone before bed, there are ways that you can do so and still get a good night's sleep. Wearing special glasses with amber-coloured lenses will filter out blue, low-wavelength light, allowing you to stare at your screen for as long as you like. Companies such as Uvex (uvex-safety.co.uk) make blue-blocking glasses and goggles in a range of styles. Alternatively, you could use computer software such as f.lux (justgetflux.com) and smartphone apps such as Twilight (play.google.com) that automatically adjusts your screen to filter out blue light between sunset and sunrise, replacing it with a softer red light.

Filter out blue light with a pair of amber-tinted glasses

©Art Agency

159

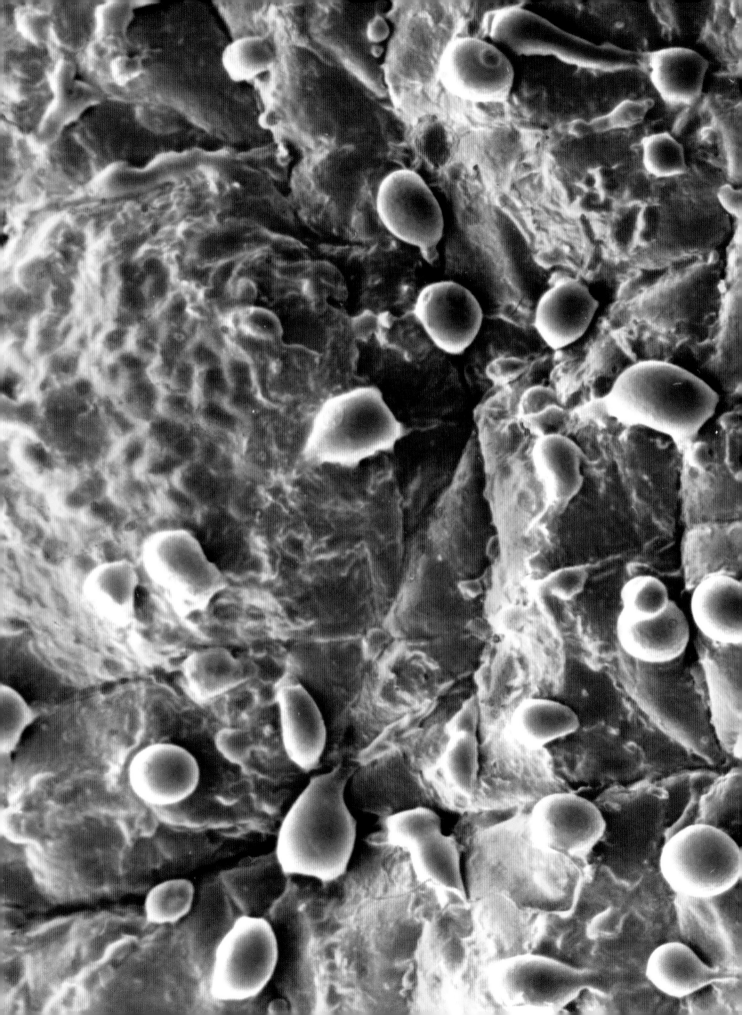

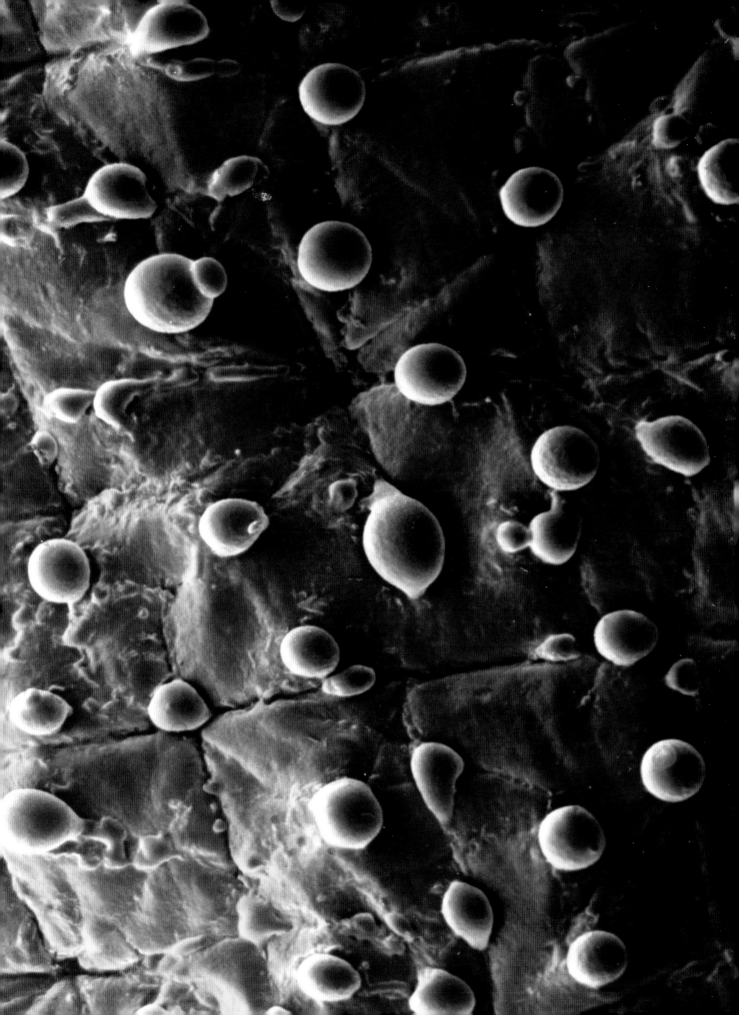

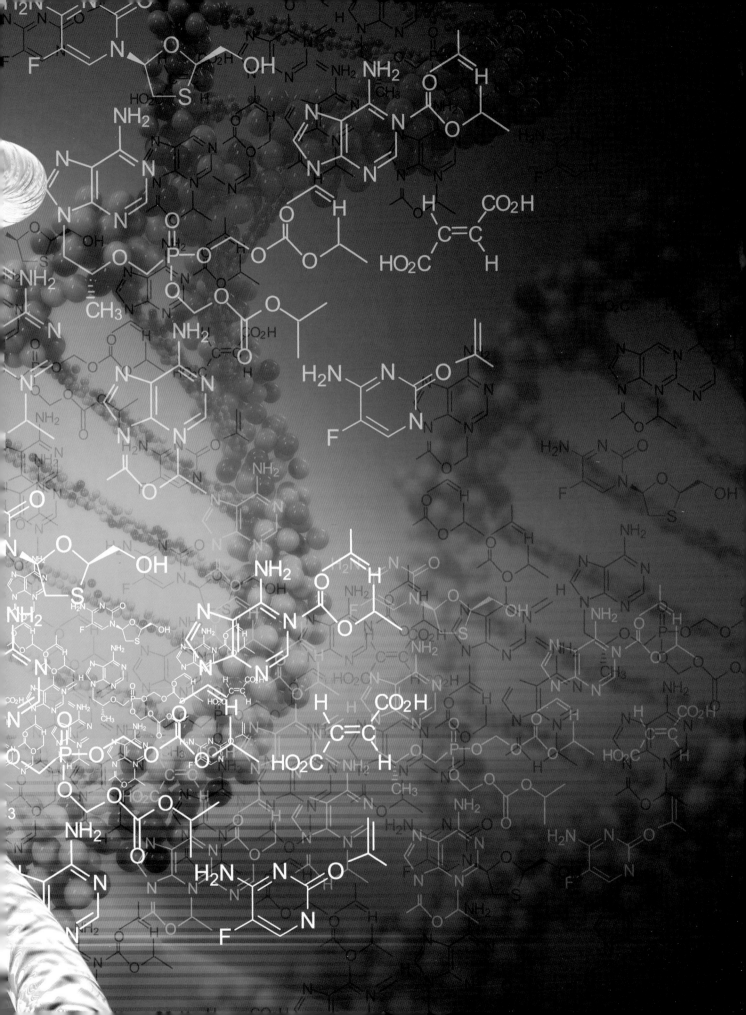